Robotics in Germany and Japan

DRESDEN PHILOSOPHY OF TECHNOLOGY STUDIES
DRESDNER STUDIEN ZUR PHILOSOPHIE DER TECHNOLOGIE

Edited by / Herausgegeben von Bernhard Irrgang

Vol./Bd. 5

Michael Funk / Bernhard Irrgang (eds.)

Robotics in Germany and Japan

Philosophical and Technical Perspectives

Bibliographic Information published by the Deutsche Nationalbibliothek
The Deutsche Nationalbibliothek lists this publication in the Deutsche Nationalbibliografie; detailed bibliographic data is available in the internet at http://dnb.d-nb.de.

Library of Congress Cataloging-in-Publication Data

Robotics in Germany and Japan : philosophical and technical perspectives / Michael Funk, Bernhard Irrgang (eds.).
pages cm — (Dresden philosophy of technology perspectives, ISSN 1861-423X ; v. 5)
ISBN 978-3-631-62071-7 — ISBN 978-3-653-03976-4 (ebook)
1. Robotics—Germany—Popular works. 2. Robotics—Japan--Popular works. 3. Robotics—Philosophy. I. Funk, Michael, 1985- editor of compilation. II. Irrgang, Bernhard, editor of compilation.
TJ211.15.R626 2014
629.8'920943—dc23

2013045885

Cover illustration:
Humanoid Robot "ARMAR" (KIT, Germany),
Photograph: Michael Funk

ISSN 1861-423X
ISBN 978-3-631-62071-7 (Print)
E-ISBN 978-3-653-03976-4 (E-Book)
DOI 10.3726/978-3-653-03976-4

Peter Lang Edition is an Imprint of Peter Lang GmbH.

Peter Lang – Frankfurt am Main · Bern · Bruxelles · New York · Oxford · Warszawa · Wien

This book is part of the Peter Lang Edition list and was peer reviewed prior to publication.

www.peterlang.com

Preface

Germany and Japan are two of the worldwide leading countries in robotics research. Robotics is a key technology and it brings about technical tasks for engineers, but also philosophical and cultural challenges. How are we going to use robots that have a human-like appearance in everyday life? What is technologically possible? What are the cultural similarities and differences between Germany and Japan? Those are some of the questions discussed in the book. Five chapters embrace an intercultural and interdisciplinary framework including current research fields like *Roboethics*, *Hermeneutics of Technologies*, *Technology Assessment*, *Robotics in Japanese Popular Culture* and *Music Robots*. Contributions on cultural interrelations, technical visions and essays round the volume off. Most of the contributions in this book are based on the lectures of the conference "Future of Robotics in Germany and Japan" (TU Dresden, November 11-12, 2010), which was made possible and kindly supported by

- *Fritz Thyssen Stiftung*

and

- *MEXT Global COE Program*: The University of Tokyo Center for Philosophy (UTCP), JSPS Kakenhi (Grants-in-Aid for Scientific Research), No. 21520004.

The editors would like to express their thankfulness! Moreover we also give our thanks to Andreas Bork, Paul Stadelhofer and Beatrix Weber most sincerely for proofreading.

Michael Funk & Bernhard Irrgang

Table of Contents

Introduction

From Fiction to Science: A German-Japanese Era-Project 11
Walther Ch. Zimmerli

Philosophical Frameworks

Robotics as a Future Vision for Hypermodern Technologies 29
Bernhard Irrgang

Roboethics and the Synthetic Approach
– A Perspective on Roboethics from Japanese Robotics Research 45
Kohji Ishihara

Robotic Appearances and Forms of Life. A Phenomenological-Hermeneutical Approach to the Relation between Robotics and Culture 59
Mark Coeckelbergh

Humanoid Robots and Human Knowing
– Perspectivity and Hermeneutics in Terms of Material Culture 69
Michael Funk

Technology Assessment

Who is taking over? Technology Assessment of Autonomous (Service) Robots 91
Michael Decker

Popular Culture and Music Robots

Robots in Japanese Popular Culture 113
Maika Nakao

Understanding the Feasibility and Applicability of the Musician-Humanoid Interaction Research: A Study of the Impression of the Musical Interaction 125
Jorge Solis & Atsuo Takanishi

Mozart to Robot – Cultural Challenges of Musical Instruments 135
Michael Funk & Jörg Jewanski

Essays

Android Robots between Service and the Apocalypse of the Human Being 147

Gerd Grübler

Joseph Weizenbaum, Responsibility and Humanoid Robots 163

Kerstin Palatini

Social Stereotypes as a Guarantee for Proper Human-Robot Interaction? Remarks to an Anthropomorphic Robot Design 171

Manja Unger-Büttner

Authors and Contact 179

Introduction

From Fiction to Science: A German-Japanese Era-Project

Walther Ch. Zimmerli

Introduction

What I am planning to do is to discuss in the first step robots as a philosophical problem. In the second step I would like to draw our attention to a previous version of that question, which was very popular when we were still dealing with the philosophical standard problems of Artificial Intelligence: the question of human minds and machines. Afterwards I would like to have a look at the transformation of the Artificial-Intelligence-discussion into the robotics-discussion and even further on beyond the notion of software, which will be decisive for my ideas. And by doing this I would finally like to focus on the question regarding cultural differences, more specifically: on the cultural differences between Asian (especially Japanese) and the European (especially the German) way of talking as well as thinking about and constructing of robots or of robotics.[1]

Before we do that, however, we have to ask ourselves: What are the problems with robotics or robots? One of the problems can even be seen reading the wordplay on the signs in the hall pointing to our "robotic conference." If it would be a conference on robots mainly or on robotics, it would read "robotics conference." But it says "robotic conference," which could result in the obviously somewhat misleading conclusion that we all here are robots, which is however, philosophically speaking, not so farfetched. Because as we all know and as I will be elaborating in what follows: within the European Philosophy the idea of artificial human beings has been inherent in our humanistic tradition since its very beginning. Even before mankind actually developed machines in the strict sense of the word, the idea of human beings as a certain kind of machines was already discussed. And that will be just one of the aspects, which we will be dealing with.

As another preliminary remark I would like to focus on the problem, whether the question concerning robots is pointing to one of the decisive characteristics of our time. To put it differently: whether it is true or not that robots will become or have already become one of the main features in the development of our society? This question includes two other questions: the question concerning robots, these little individual entities, beings, gadgets and the context in which

1 As far back as 1989 I published an article "Human Minds, Robots, and the Technician of the Future" (Zimmerli 1989). At that time the AI-discussion as well as the debate on robots was almost a part of the philosophy of mind. Today, however, the discourse on both AI and robotization are strongly influenced by the network-paradigm and the subsequent ideas on globalization. Therefore the deliberations in this article are to be considered as a revision of my previous ideas in the light of the overwhelming omnipresence of a globalized network, i.e. the WWW.

they interact with human beings on the one hand and robotics in the perspective of an academic discipline on the other.

So the question I will begin with reads: "Is it true that robots will become or have already become one of the main features in the development of our society?" And everyone who is philosophically trained knows, of course, that here we are running into some kind of an immanently contradiction. It is a contradiction between true and will. As is well known this is an Aristotelian problem. How do we decide on the truth value of a proposition in the future tense? Since Aristotle we know that this is impossible. We do not have any way of deciding, whether a proposition in the future tense is true or not, unless it is changing from a proposition about the future into a descriptive proposition about the presence (cf. Aristotle, Perihermeneias, 18b ff.). So what we are actually asking when we ask: "Is it true that robots will become one of the main features in the development of our society?" is: "Is it a valid hypothesis that robots and robotics will become or rather have already become a decisive main feature of our present?" And keeping in mind that this is the main question I would like to proceed now to the first step.

Robots as a Philosophical Problem

Robots: The Vision

As is well known the problem with robots is already a problem in Ancient Greek Philosophy. The "locus classicus" is to be found in Aristotle´s "Politics" and it reads, in the hypothetical way of predictions, as follows:

> "For if every instrument could accomplish its own work or obeying or anticipating the will of others, like the statues of Daedalus, or the tripods of Hephaestus, which, says the poet, 'of their own accord entered the assembly of the Gods;' if, in like manner, the shuttle would weave and the plectrum touch the lyre without a hand to guide them, chief workmen would not want servants, nor masters slaves." (Aristotle, Politics, Book One, Part IV, 53b)

That is a gorgeous vision indeed, as it has been written by someone who did not even know mechanical instruments, not to mention automats. A person who just from the very imagination of his mastermind envisioned the possibility that if these kind of automatic instruments would already be in place then we would have reached, as Marx and Engels has put it, "a society without classes" (cf. Marx & Engels 1943): no servants, no masters. Now if we keep in mind that Aristotle was envisioning a future like this then we can easily see that he is not talking about robots, because the very notion of "robot" is Russian, and Aristotle of course, did not know Russian. Therefore he did not talk about robots, but by definition he did so nonetheless by talking about mechanical instruments displaying intelligent behavior and that is the definition of robots I will be starting from.

Later on we will learn that we have to distinguish between robots and *meta-robots*. By *meta-robots* we understand systems consisting of mechanical instruments displaying intelligent behavior beyond the intelligent behavior of the individual mechanical instrument. As always Aristotle was the first to offer some kind of definition. As indicated above, in the early modern times European Philosophy focused on the problem, whether we can distinguish between observing intelligent behavior or not. We are, however, capable of distinguishing a human being displaying intelligent behavior from a machine displaying intelligent behavior. So the question is: Are we human beings or just machines displaying intelligent behavior?

The idea of the "automaton spiritual" – formulated by the very same notion used by Descartes, Spinonza and Leibniz (cf. Descartes 1662/1984; Spinoza 1661/2003; Leibnitz 1854, 61 et passim; see also Lohen 1966) – implies that we cannot really tell, whether we observe the intelligent behavior of a machine or intelligent behavior of human beings, if we just observe intelligent behavior. And of course, in the period of Enlightenment the idea of "L´Homme Machine" (cf. La Mettrie 1996), of man as a machine was quite common at least among the Materialistic Philosophy. And from then on, this notion of the distinguishability or non-distinguishability of human beings displaying intelligent behavior and machines displaying intelligent behavior has become one of the key questions within what later has been called the "Philosophy of Artificial Intelligence."[2] But before we talk about that let us return to Aristotle who said as already mentioned, "chief workmen would not want servants, nor masters slaves" (Aristotle, Politics, loc. cit.). Robotics, robots or machines displaying intelligent behavior have something to do with what we call "labor."

The Anthropological Dimension

Accordingly, the idea of robots is in some way from the very beginning connected to the idea of labor. If we now look at the anthropological dimension, it is not farfetched to draw the following conclusions:

- First: If human beings are defined by labor, labor could be understood as the self-objectivation of the internal human nature by changing the external nature (you could also put it more briefly: if human beings are defined by technology, that is by changing external nature while self-objectifying the human internal nature, such like ideas or concepts)
- Second: If human labor is being partly performed by robots, then a robotic world or a completely robotized world would be a world deprived of human nature. This would be the anthropological problem inherent in robots` nature,

2 Some of the most important philosophical texts in the beginning debate on Artificial Intelligence have been collected, translated into German and commented by Zimmerli & Wolf (2002).

if robots, machines or mechanical devices would indeed display behavior and by doing so change the external human nature or technologically altering external nature.

The Economic Dimension

If we look at the economic dimension of talking about labor, we can see another possible conclusion: If the objective of human labor is value creation, both economic and ideal, human labor is not just being defined by creating value for the market, but also by creating value for self-esteem and self-realization. Just think about unemployed people. With unemployed people the problem is not that most of them do not make enough money, because depending on the kind of social security system they get quite a decent amount of money nevertheless. The problem is that they are being deprived of value creation, of contributing in a more or less ideal sense to the value creation of the society. And if values in the strict realistic sense are depending on human impact (if humans are value creating beings and if values themselves are dependent on human impact) then a completely robotized world would not create values in the strict sense of the word.

Of course, we know that there will never be a completely robotized world. But nevertheless we have to take these "counterfactual conditionals" into consideration: If there would be a completely robotized world then we would not have value creation in the strict sense of the word. And then we would not have human labor in the strict sense of the word either. Thus we keep in mind: the anthropological and economic counterfactual conditionals are dealing with a completely robotized world.

Human Minds and Machines

Artificial Intelligence (AI) and Robots

If we look at the topic of the relation of robotization to the creation of values, to the labor force and (as we will see later) to industry, which of course is not a case of counterfactual conditionals, then we could remind ourselves of the history of Artificial Intelligence (AI) and robots. As I tried to point out, both the notion of "robots" and the notion of the "artificial technologically intelligent devices (machines)" are inherent in the philosophy at least since the beginning of Modern Times or as we have seen in some respects even since Ancient Times. More than 25 years ago this seems to have been the main problem of robots and robotization: Namely, that,

> "the industrial manipulating equipment (machines), erroneously called robots in common parlance, takes, in the pros and cons of the discussion, a position which objectively speaking, they should not take. This is the fault of the term 'robot,' which due to its use in science-fiction literature and films, gives the wrong impression of quasi-intelligent beings. As long as this misunderstanding exists or is even encouraged by the use of this term, it will be very difficult to maintain an objective discussion." (Honrath 1984, p. 16 [Translated by the author, Walther Ch. Zimmerli.])

The sociologist Honrath, who formulated this critical idea, was quite right by focusing on a misinterpretation or a misconception of the notion "robots." If you mistake a robot for an "android," for a quasi-human being or a mechanical human being, then you will never find a way out of this problem. You will actually never find a solution to the question, whether and in which respect a system, which uses robots, can be a system useful for mankind or for human society. A meaningful notion of "robot" and "robotization" would not be this kind of "android" idea. But it could boil down to the automatization of the production process in industry. Of course, we could not deal with all the different other aspects of robotization and robots. But in the following section of my deliberations, I will focus on robotization as automatization of the production process in industry.

We know that the very notion of "industry" is ambivalent in itself, because we know that we tend to think about our society as a "post-industrial society" (cf. Bell 1973). But on the other hand we know that this idea of a post-industrial society can be developed within the framework of an industrial society only. So although we are still living in an industrial society we are within this industrial society envisaging the idea of a post-industrial society, but nevertheless our economy mainly relies on industry production, as we can see especially after the worldwide financial crisis. That is why Germany is doing relatively well these days, because Germany still is an industry nation and is living of exports of industrial products, especially of cars. Just think of Volkswagen, a global player automobile company which right now and in the years to come it will probably be the number one worldwide as far as the numbers of cars produced and as far as the economic value of the company is concerned.

As you might recall, Artificial Intelligence was not from the very beginning of the discussion connected to the topic of robots only. The debate on Artificial Intelligence at its very beginning since the 1950s of the last century was primarily just dealing with computers, not with robots, not with technological devices displaying intelligent mechanical behavior, but with machines displaying on their screens language, i.e. intelligent linguistic behavior. The intriguing question was: Is it possible that machines can think? (cf. Turing 1959) It was not the question: Is it possible, that machines can display intelligent labor or behavior, but: is it possible that machines can *think*? Having dealt with these questions quote extensively, now 60 years later, the problem seems to be solved. Of course, machines are capable of displaying intelligent behavior, both in a language connected and an in a mechanical way. Thus the idea of robots, which I have been discussing above, is one aspect of the problem of artificial intelligence. The other aspect of course is the question: How intelligent can a machine be? Or: Which aspects of human intelligent behavior can be simulated or maybe even optimized by "intelligent" machines?

Distributed Artificial Intelligence (DAI) or Robots?

The next step of the discussion on robots, however, was focusing on Distributed Artificial Intelligence (DAI), i.e. on connecting robots, capable of carrying out the functions necessary in industry. Distributed Artificial Intelligence from its very beginning in the late 1980s and early 1990s of the last century was not focused on simulating chess playing or other behavior of human beings, which could be called "intelligent," but were focused towards the industrial production system. Looking at the industrial production system we can divide the human or non-human activities within the industrial system into two different parts, namely on the one hand dealing with innovation and optimization or as we usually call it "Research and Development" (R&D); and on the other hand dealing with production itself. Our leading question therefore has to be reformulated: "Which one of (these two) functions in industry could also be carried out by robots, i.e. mechanical devices displaying 'intelligent' behavior?" And the answer, of course, was in the discourse of DAI: Just the production! Innovation and optimization are still depending on human creativity and market integration. Thus innovation cannot be completely simulated by machine intelligence. That was the state of the discussion beginning in the 1990s.

The Concept of Networks

In those days we did not have the World Wide Web (www) yet. Of course another world wide web called "ARPAnet" ("Advanced Research Project Agency"), a network between five main frame computers in five U.S. universities that had been developed in the late 1960s, did already exist (Rheingold 1993, pp. 24-26). The idea was to strengthen the defense system of the United States in order to prevent the Soviet Union from disenabling U.S. American defense forces by one massive first nuclear strike. The idea was not to build a second strike capability, that was not the strategy, but to decentralize the U.S. American defense forces` intelligence. By following this idea they just tried to decentralize the same information processing capacity in different places. If the Soviet Union would wipe out one of the centers of intelligence of the defense system, there would be still four more in order to survive. That was the beginning of the World Wide Web, following, as everybody well read in the Ancient Philosophy knows, the Pre-Socratic Philosophy of Heraclitus: "The war is the father of everything." The "father" of our civil World Wide Web, of the world wide system of information exchange, was a military system. And the idea behind it was to distribute interlinked computer systems and thus decentralize them.

In the beginning 1990s, however, some computer freaks started to do something very simple: they started to connect their individual computers to the telephone-net. From that very moment on, the world of the World Wide Web started. It took roughly ten years of pioneer work and these pioneers did exactly this, before the idea of the net itself was being exploited (Rheingold 1993, pp. 24-26).

Why was this net the next step after Artificial Intelligence (AI) and Distributed Artificial Intelligence (DAI)? Well, because all the intelligent actions on the individual computers (or the individual robots) were much more secure and much more flexible in a decentralized structure. And because you have this plasticity of the whole system, if you connect a lot of individual devices displaying intelligent behavior and creating in this way *meta-intelligence* (the system itself, the net), you get a lot of centers distributed among the different nodes of the net. In an individual computer there is a CPU, the Central Processing Unit, and then there are all the other peripheral devices. Whereas in the net there is no Central Operating Unit and each node in the network is in principle capable of functioning as a CPU. That is one of the advantages. And of course, if there are nodes, there are *meta-nodes* and different operating systems. And as long as you are in the possession of the power of regulating the processing system (as long as you are called "Bill Gates") you are in power. The most powerful men of the world in those days were not the presidents of the United States or the Soviet Union. The most powerful men were the men, who owned these centralized units, which were capable of running the whole rest of the system. And that was the basic idea of a technical network.

Since the 1990s we know that this idea of a network has become a paradigm and even a *meta-paradigm*, i.e. a paradigm which is applicable in all different realms of industry, of science, of society etc. And if we pay more attention to the problem of the roots of this idea then it becomes obvious that it is something getting ever more prominent since the 1960s (Zimmerli 1998). Not technical networks, but the idea of networks literally popped off the ground in all different disciplines including Artificial Intelligence. For example the concept of memory as a semantic network or the idea of network as the main structure behind all different actors in economy goes back to the 1960s. And in the social sciences, the idea of network as a basic concept of all social systems was developed at the same time. In brief: from the 1960s on you will find this idea of networks as some kind of an overarching *meta-paradigm*, increasingly applied all over academia, both in computer sciences and other academic disciplines.

Beyond Software

Production of Software by Machines?

After the idea of network became a *meta-paradigm* in all different disciplines, in philosophy and science as well as in technology it had been realized by the creation of the World Wide Web. The concept of network is not just a heuristic idea, but an idea which has empirical existence. From that point on, the scientific and the technological community started to understand the idea of a human-machine or a man-machine hybrid. It is not just the idea of computers being connected. Because computers are being connected literally do nothing, unless they are part of human-machine-systems, i.e. unless human beings are involved. If one looks for instance at the screen of a PC then one sees changes of physical states. But in order to interpret these as different letters or different signs, in order to interpret signs at all, one needs a human observer that is also a human user. The machine-user entity is the node, not the node of the technical network as such.

The hybrids of human beings and machines are connected to networks. And therefore, the hybrid networks (the human-machine-networks) replace the idea of individual artificial intelligent entities such like computers or robots. From the 1990s on we did not talk anymore about robots as individual intelligent entities or individual mechanical entities displaying intelligent behavior. Now we are talking about systems, networks of mechanical devices and human beings displaying intelligent behavior together in processes of interaction. Again the idea began to develop – like 50 years before – that there is a decisive difference between hardware and software, between the mechanical device (be it electrical, chemical or whatsoever) and the program running on that mechanical device. Our wet-wear which we have in our skulls, our brains is of course also entities on which programs, i.e. software is running. There is an interesting development to be observed when we project this back into history. By doing this it becomes obvious that the definition of the humankind of human beings was always a mirror of the most advanced pieces of technology.

At the beginning of the Modern Age in the mechanical thinking human beings of course were considered to be complicated clocks. Just think of Leibniz, for whom of course the whole world was a clock work and God was the clock maker, who had manufactured all these perfectly running clocks. The idea of pre-stabilized harmony is illustrated by Leibniz with the image of a clockmaker, who created absolutely synchronized clocks (Bayle 1978, 86b). And then we can see it happen all over again, when in the 19th-century the human brain was perceived to be something like a very complicated telephone system (cf. Searle 1984). And since the 1950s and 1960s the neurosciences perceived the human brain as a computer (cf. Anderson & Ross 1964). From the 1980s on the idea of self-organizing networks, neural networks replaced the idea of the human brain being a computer (cf. Hopfield 1988; Churchland 1986). And almost the same is happening now again, if the step from the individual mechanical technical devices displaying intelligent behavior is made to the system of the coordination of individual mechanical devices displaying intelligent behavior (cf. Beni 2004).

It is in this context that the old question of the individual robots shows up again with respect to software:

- Could it be possible for us to design devices which would be capable of designing other devices, could we e.g. design a robot that could be capable of designing other robots? Can machines or computers produce software?

And the answer of course is highly disputed: Yes, it is no longer impossible, it is rather easily possible. So the next question comes up:

- Can machines or computers distinguish between software, which is suitable for the realization of a set aim and other software? And if so, can intelligent machines distinguish between valuable and less valuable software within mechanical devices displaying intelligent behavior? Or to put it differently: Is a robot capable of distinguishing between robots, which are useful for a set aim and others, which are not useful for a set aim?

And the answer is again: Yes, of course is it possible. We can even talk about quality assurance. We can even design systems in which the whole quality assurance of the system is performed by the intelligent system itself or by a redundant second one. And now the decisive question comes up:

- Can machines or computers offer nominal values that include an aim towards the realization, of which the software judged to be suitable used? In other words: Can machines deal with values?

If that is the case then of course the main anthropological issue would be solved by the dissolution of the main anthropological issue. There would be no privileged situation for human beings anymore, if robots could do the same. So the answer to this question is: No. I should put it differently: the answer has to be: No. If we try to stick to the notion of a "human being" and if we try to stick to the notion of "human-machine-hybrid-systems" then there should be at least one decisive difference between human beings and machines displaying intelligent behavior. And this decisive difference is probably the capability of defining and setting values. Consequently the answer is "No" and has to be "No." If the answer would be "Yes," then we would have to start a new round of deliberations. For instance, if we would apply this to the production system as I said above.

The production system as such is a network. Since the beginning of industrial production we followed a different paradigm, not yet the idea of network that was a *meta-paradigm*, which as I said before came into being in the middle of the 20th-century only. The beginning industrial production followed the paradigm of the chain, the "great chain of being" (Lovejoy 1936), and that goes back to the 16th-century. The idea of the great chain, i.e. a sequence of individual actions, or of individual entities, or of individual beings, or of individual elements was the prevalent idea and is still very much alive. If you look at the problems for instance, which we have not dealing with robots (at least in the European context) but dealing with Bologna, then you see the main problem: that the idea of a linear study course or curriculum (in German: "Studiengang") is replaced by the idea of a modularized network. There is no set sequence that one needs to have the first step at first and the second step afterwards. But the idea is that it is possible to combine all these different modules in different ways. The sequence, the great chain of being is replaced by a great modularized network of being.

Assembly Line Robots

Looking at the production system we see that the dominant idea, even if we are in actual fact already working within a network is still one of a sequence or a line: the assembly line. If we look at late modern production systems we see, however, that it is not organized according to the tradition of the assembly line anymore, but rather according to an integrated network of different assembly lines (and the same applies to the traditional idea of supply chains) (cf. Zimmerli, Bagusat & Müller 2007). But the main idea is nevertheless still that of an assembly line of robots, the type of robots e.g. you usually see at car production. They do not show "android" features, they do not look homo erectus-like, but you see e.g. one great arm. And this great arm performs almost the same movements all the time. That is what assembly line robots are all about. They seem to move on their own free will. One looks at them and they seem to display intelligent autonomous behavior. And one would not even be capable to tell whether they are actually internally programmed to do that or whether they are developing their program as they go along. It would not be possible to tell that because there is an asymmetry between the possibility to explain these movements after the fact and the possibility to predict them before, i.e. an asymmetry between explanation and prediction. These assembly line robots seem to move on their own free will. Under certain circumstances and under different conditions of carrying out various operations, they may seem (to some observers) like independent beings, who program their own set values.

What we have seen before applies here too: if the only action which these robots cannot do, the only limits to robots taking over work operations in the production process would be the setting of the values by themselves, then of course everything else, every operation that can be described by software production analysis methods can be carried out by every suitably implemented machine. The only thing that cannot be carried out by suitably implemented machines is the setting of the set values itself. And if we apply the notion of "network," which we have developed in our last step, then it becomes obvious that the operation of networks of robots can easily be programmed, be it by human beings or by machines: Of course we can design machines, which are capable of designing networks of programs of machines.

Beyond Swarm-Intelligence

And that is now where we enter the next level of the Artificial Intelligence discussion, the debate on the so called "swarm-intelligence." Originally swarm-intelligence used to be a notion taken from the biological sphere, especially with respect to the behavior of ants, birds and fish. But these days we speak about swarm-intelligence of robots, especially of nano-robots, which are very tiny little devices displaying intelligent behavior themselves (cf. Winfield, Harper & Nembrini 2005). Or to put it more precisely: the behavior displayed by swarms of small robots ("nanobots") is not a behavior called "intelligent," because the individual nanobots display intelligent behavior, but only because the system of intelligent nanobots is displaying intelligent behavior. That is why the very notion of "robots programmed in an assembly line (or any other kind of collaborative pattern)" is presupposing the idea of a decentralized *meta-program* or rather a behavioral collaborative pattern of robots (or other "intelligent" machines) – and this idea is what we call "swarm-intelligence."

So the question is not whether the individual robot in an assembly line performs intelligent behavior (that is always the case, because that is what they are supposed to do) but the assembly line itself has to display intelligent behavior. Example: given that a robot as the individual intelligent production machine always takes a piece or device and puts it over here and does all the time with it, what it is programmed to do, and given that there is no piece or device around then the whole system does not work. So the individual robot could perform the same operation as many times as it is programmed, but if the supply of the elements needed to perform an actual labor action by doing this would be lacking, i.e. if the system would not be so efficient by not having a sufficient supply of elements or devices, then the system itself would not be behaving in an intelligent way; it would run empty.

So far the paradigmatic metaphor that dominated the world in general and the individual production in special was that of "the great chain of being" as pointed out above, and within the world of production it was the notion of "supply chain" but if we look at the chains we have, e.g. at the assembly line and the supply chain, we see two chains crossing each other and by doing so forming a node of a network. If the network itself is not programmed intelligently then there is no way of talking about intelligent behavior of robots.

Therefore, we have to deal from now on with what I call *meta-programs* or *meta-robots*. The discourse on robots has in actual fact already been replaced by the discourse on *meta-robots*, if we talk about robots in this new system-oriented way. With regard to this *meta-robot*, however, an asymmetry-hypothesis (that seems to be valid) between observation of behavior and programming applies. Although we can observe intelligent behavior it is possible that we cannot program it. What we find here is the next level of the asymmetry between explanation and prediction in the philosophy of science and it is called the difference between programming and teaching (cf. Poprawe 2005, pp. 376-378). The very notion of "teaching" is used here in a system, which does not deal with the interaction between teachers and students at all, but with the interaction between human beings and machines. For instance, if one tries to program a laser welding device then one does not have to write programs anymore. These machines do have written programs, of course. But these are written programs of learning. Consequently, what we do is teaching the machine by actually physically guiding it, by taking the laser welding arm and putting it into some place and then putting it over and bringing it back, and then putting it back over there. And after doing that a few times the machine has "learned" it, and from then on it does it by itself. Of course, it does not literally do it by itself; it is not autonomous in the very strict philosophical sense of "autonomy." But we have taught the machine to do that and the machine does it, unless we teach it to behave differently. We can teach the machine by talking to it, in written language (program) or by showing it how to behave, by taking the laser welding arm again and putting it to a different place etc. By doing so a few times, you have taught the machine to behave like an intelligent being. And pretty much the same applies to swarm-intelligence of a system which has to be "taught" and not just programmed.

And still another analogy becomes heuristically viable. In the traditional AI-debate we did distinguish between a strong artificial intelligence notion, which was applied to the machine itself, and a weak notion of artificial intelligence, which was defined by the machine behaving *as if* it were intelligent (Searle op. cit, cf.; Zimmerli & Wolf 2002, p. 21). And quite analogously we now can distinguish between a strong and a weak notion of *meta-robots*. The strong notion of *meta-robots* would be that the whole system of human-machine-elements behaving in an intelligent way is itself intelligent: the network, the production house, the fabric is intelligent. The weak notion, however, would consist in the claim that it behaves *as if* it were intelligent, as if it were itself an intelligent entity.

Cultural Differences

Identity of Technology – Differences in Culture

What I have analyzed so far is pretty much the same all over the world, as mathematics and physics is pretty much the same all over the world. So this kind of *meta-robotics* (as a notion for the academic discipline) is the same all over the world. Nevertheless, there are differences, especially with respect to the labor force and to the style of production. What I am saying is: Although there is an identity of technology, there are differences in culture. These are just analytical terms, because they apply to an everyday non-technical notion of "technology and culture" only. Because if we look at the technical (or terminological) definition of "culture" and of "technology" we can easily see that technology in actual fact *is* culture (cf. Zimmerli 2005). Therefore, what I am dealing with now is the ordinary language difference between "technology" and "culture." If we compare these *meta-robots* – and again we have to bear in mind that *meta-robots* is the notion of human-machine-networks in different parts of the world – e.g. with respect to car manufacturing, then we come up with the following conclusion: The machine parts (the robots in the very sense of the word) are identical. They are produced in Japan or in Germany or in the U.S., and they are built into the whole quasi intelligent robot system. The *meta-robots*, the assembly line programs themselves, however, might be different, because there are different ways or interpretation patterns or style, e.g. the Toyota-Way of putting together elements into cars which was very famous in the 1990s (Liker 2003). And there was (and is) the Volkswagen-Way, which was quite different, and rather oriented towards the traditional assembly line type of production, but for the time being rather more successful than the Toyota way.

Nature, Technology and Culture

Nevertheless, if we now try to remind ourselves that technology is defined as the way the internal human nature shapes the external nature and if the different ways of changing nature by human beings are called "culture," then it goes without saying that the different cultures define the different ways of technology. On this *meta-robot*-level we are not interested in the technical aspect alone, but in the human-machine-system. Thus the question is how the different cultures are of importance with respect to the different human-machine-systems and how they in turn might influence both the different cultures and the different ways of technologically shaping the external nature.

Stereotypes: European Individuality and Asian Collectivity

What we therefore have to do now is to identify the differences in culture with respect to robotics, and in doing so we immediately are confronted with general assumptions about the (national) characters of different civilizations. These general assumptions we usually call "stereotypes" (cf. Cox et al. 2012). To give you some examples, we come to those stereotypes, which we usually apply when we talk about different cultures. What is the difference between Asian cultures, especially the Japanese, and European ones, especially the German culture with respect to the production system, e.g. to the assembly line organization?

The most prominent distinction of stereotypes in this respect is the distinction between European and Asian behavior: Europeans favoring individuality – and there is much historical evidence for that –, and Asians are favoring collectivity – and there is much evidence for that too. But if we take a closer look it becomes quite obvious that since globalization has started this is changing, and it is no coincidence that globalization started approximately at the time when the network paradigm became real by the World Wide Web, because there is no globalization except for the real time action, the possibility of performing synchronous action in real time. And that again presupposes the network, the World Wide Web. Globalization, however, (including a universally unifying technology) results in an (at least experimental) mutual exchange of cultures of production. If we look for instance at different cultures of production, which affect robots in Japan and in Germany, and just focus on the car industry again, then we see both different cultures experimentally test the other system. Or to put it into a nutshell: Kaizen as the core-idea of the production philosophy in the Japanese way of production, as well as Taylorism as the philosophy behind the assembly line system of the European car production are nowadays practiced by all robotized "Original Equipment Manufacturers" ("OEMs"), in this case by Toyota and Volkswagen.

Globalization seems to have resulted in a possible mutual exchange of the identities of the cultures behind the different *meta-robots* or robot systems. Consequently at the end of the day the difference of cultures is not as important any more. We do have different cultures, we know that. We do have different everyday cultures and we do have different high cultures. And if we look at the way the robot systems were implemented, then we see there were cultural differences as well. But these days, at least experimentally, they are mutually interchangeable. In some respect, Volkswagen today is more Toyota than Toyota ever was, and in another respect, Toyota is more Volkswagen today than Volkswagen ever was. And this is true, even if we consider the necessity for all these big global player production companies to engage in a better understanding of the value of labor and the labor system of the other culture. For instance, Toyota and Volkswagen are producing their cars for the continent of Africa in Port Elizabeth (South Africa). So the labor force is African. They have to adapt to the African culture of labor and of production. And they do that by applying the very same *meta-robot*-systems in South Africa as they apply in Japan or in Germany.

Concluding Remarks

What I am saying is that globalization in some way indicates the end of the idea that there is a decisive cultural difference between robotization in different cultures. And that does not come as a surprise. If you would have had to guess at the beginning, which would be the answer to the question I was putting, namely: Whether robotization would become one of the main decisive aspects of our everyday culture, you probably would have guessed exactly that. But now we do not have to guess anymore; now we do not only know it, but we even know the reasons.

References

Anderson, J. A. & A. Ross (eds.) 1964: Minds and Machines, Englewood Cliffs/N. J.

Bayle, P. 1978: "Art. Rorarius. " In: Historisches und Christliches Wörterbuch, Leipzig 1744, Nachdruck Hildesheim & New York 1978, 86b.

Bell, D. 1973: The Coming of the Post-Industrial Society, Frankfurt am Main.

Beni, G. 2004: "From Swarm Intelligence to Swarm Robotics." In: Sahin, E. & W. M. Spears (eds.) 2004: Swarm Robotics, SAB 2004 International Workshop Santa Monica, CA.

Churchland, P. 1986: Neurophilosophy. Toward a Unified Science of the Mind-Brain, Cambridge.

Cox, W. T. L., L. Y. Abramson, P. G. Devine & S. D. Hollon 2012: "Stereotypes, Prejudice, and Depression: The Integrated Perspective." In: Perspectives on Psychological Science, 7 (5), pp. 427-449.

Descartes R. 1984: Traitè de l`Homme (1662), Hildesheim.

Honrath, K. (ed.) 1984: Schritte zur automatisierten Produktion. Zukunftsorientierte Konzepte für den mittleren Maschinenbau am Beispiel realisierter Lösungen. VDI-Ges. Prod.-

Technik, Düsseldorf.
Hopfield, J. J. 1988: "Neural Networks and Physical Systems with Emergent Collective Computational Abilities." In: Anderson J. A. &. E. Rosenfeld (eds.) 1988: Neurocomputing. Foundations of Research, Cambridge.
La Mettrie, J. O. de 1996: "Machine Man." In: Machine Man and other Writings, Cambridge, pp. 1-40.
Leibniz, G. W. 1854: Réfutations Inédites de Spinonza par Leibniz. ed. by Foucher de Careil, Paris.
Liker, J. 2003: The Toyota-Way: 14 Management Principles from the World`s Greatest Manufacturer, New York.
Lohen , L. 1966: Human Robots in Myth and Science, London.
Lovejoy, A. O. 1936: The Great Chain of Being: A Study of the History of an Idea, Cambridge (MA).
Marx, K. & F. Engels 1943: Manifesto of the Communist Party. Selected Works. Vol. 1, London.
Poprawe, R. 2005: Lasertechnik für die Fertigung – Grundlagen, Perspektiven und Beispiele für den innovativen Ingenieur, Berlin.
Rheingold, H. 1993: The Virtual Community. Homesteading on the Electronic Frontier, Reading (Mass.).
Searle, J. 1984: Minds, Brains and Science, London.
Spinoza, B. 2003: Tractatus de Intellectus Emendatione (1661), Paris.
Turing, A. 1959: "Computing Machinery and Intelligence." In: Mind, 59, pp. 433-460.
Winfield, A. F.T., C. J. Harper & J. Nembrini 2004: "Towards Dependable Swarm and a New Discipline of Swarm Engineering." In: Sahin, E.& W. M. Separs (eds.) 2004: Swarm Robotics. SAB 2004. International Workshop Santa Monica. CA. USA. July 2004. Revisited Selected Papers, Berlin 2005, pp. 126-142.
Zimmerli, W. Ch. 1989: "Human Minds, Robots, and the Technician of the Future." In: Research in Philosophy & Technology, Vol. 9 (1989), pp. 183-196.
Zimmerli, W. Ch. 1998: "The Context: Virtuality and Networking after Postmodernism." In: Theron, F., A. van Rooyen & F. Uys (eds.) 1998: Spanning the Global Divide. Networking for Sustainable Delivery. School of Public Management, University of Stellenbosch, pp. 1-17.
Zimmerli, W. Ch. & S. Wolf (eds.) 2002: Künstliche Intelligenz. Philosophische Probleme. 2nd ed., Stuttgart.
Zimmerli, W. Ch. 2005: Technologie als Kultur. 2nd ed., Hildesheim.
Zimmerli W. Ch., O. Bagusat & A. Müller 2007: "Bildung als Instrument eines globalen Integrationsmanagements." In: Garcia Sanz, F. J., K. Semmler & J. Walther (eds.) 2007: Die Automobilindustrie auf dem Weg zur globalen Netzwerkkompetenz, Berlin, pp. 77-89.

Philosophical Frameworks

Robotics as a Future Vision for Hypermodern Technologies

Bernhard Irrgang

The robots of today are smart, but they are not smart enough. They have to act under pressure: Their employers always require higher and more complex performance. There is especially one thing of what they have to be capable of: thinking. The almost unstoppable triumphal procession of the working machine ended up in stagnation during the past years. Even despite remarkable technical improvements, most of such systems are still comparatively dumb. For example, Car-O-bot can easily open a room door. But once the door is stocked, because the frame has got distorted, the machine becomes helpless. The same thing happens to a welding robot once the assembly line stops. Robots are not yet useful for practice. Robotics-research was focussed far too long on the necessities of only a few industries. At that time people did not invest into the intelligence of robots, but into optimizing the environment. Now there are robots, built especially in order to be integrated into small and middle production processes. This new generation of robots is at least limitedly able to cooperate directly with human colleagues. But to enable such an improvement, the machines are required to perceive their environment, e.g. tools or instruments. One of the most important tasks is the precise proportioning of power. The dialogue between human being and machine is hard working as well (Technology Review 4/2010, pp. 58-60). But things that work in sterile and clean laboratories do not have to work on the outdoors. More and more objects and characteristics need to be included by the machine if its environment gets more and more complex. Increasing speed and stability are lowering precision and controllability. This is a cultural breakdown for the engineers, who usually try to control everything with precision (Technology Review 4/2010, pp. 62-63).

Technical Artefacts and Technical Practice: The Denotation of the Human-Machine-Interface for Robotics

The terms of "technique" and "technology" are used quite similar within the USA. Relating to my phenomenological-hermeneutical method, the following distinction is recommended: Technique refers to technical abilities and the produced artefacts as well as their use. Technology describes the technical knowledge and the teaching of technical knowledge (about technical courses of action and about operational sequences), and further the out coming machines and technological structures. Both kinds are interacting with each other and exist nowadays next to each other. I will call the sum of techniques and technologies the "technical world." An adequate definition of technique, which I, for further valuation, take as a basis even for the interaction of humans and machines, is demonstrated in the description of the production cycle. It has got three aspects with two sides in each case: It includes (1) the construction and the production of instruments and works, as well as the use of instruments, (2) the use and the consumption of technical works and (3) the disposal and the recycling of technical works and of new products in terms of a closure of the circuitry. The two aspects are (1) producing and using in terms of human actions and (2) technical instruments in order to perceive works as describable physical, chemical and biological operations. The Production of something is technical action in a proper sense, to use something refers to a determined technical handling and to dispose of something refers to technical handling in a classical meaning. The subject of Philosophy of Technology is technical and technical determined handling, since technical handling is performed to finally enable technical determined handling. The subject of Philosophy of Technology is in the strict sense the mutual relation of technical handling and technical determined handling. Different technical potentials have to be seen as a result of the connection between technical handling on the one hand and technical determined handling on the other hand. The cycle of technical and technical determined handling, as different kinds of "know how," implies a dynamization of the concept of technical handling and a transformation of the different kinds of technical practice. The classical concept of *poiesis* and the *instrumental rationality* are no longer appropriate to the current technical reality.

The human handling of a technical artefact puts it into an anthropological context, into an anthropological potential. The human being is in possession of a handling knowledge, an implicit know how of the effect, which can be achieved by technical handling. This implicit knowledge occurs as individual knowledge within individuals, even in animals (e.g. the gull and the clam), but it does not occur as a systematic world knowledge. The world knowledge is the collected handling knowledge within the world including the own finiteness. The capability of human handling is related to the world, full of theories and sense related, so by no means without world knowledge and orientated on comprehension. The human handling knowledge has always at the same time a creative character, a universal connection exceeding particular individual acting in a twofold way: On the one hand back to yourself and on the other hand exceeding onto the hidden world which might be one of the other causes for a perceived phenomenon.

Technical development can be understood as the self-organization of an interaction with technical instruments without any controlling and ruling instance and without any regulating centre or system. Technology itself does neither enforce any regulating (governmental) power, even though there are technical forces, especially in larger technology (large technical systems), which may require a governmental regulation. The terms of technical potential and technical dispositive are paraphrasing patterns of action by technical instruments (technical artefacts or nature processes), while the patterns of handling of technical practice (which was leading within the handcraft paradigm) are barely restricted to a process sequence that can be implied into machines. The three-dimensionality of technical practice is translated into a one-dimensional of technical processes within machines, if – and only if – it is implemented into machines. We could therefore speak of the three-dimensionality of the technical handler (user) and the one-dimensionality of the robot. The critical theory proofs to be true once the engineer, the "*homo faber*," understands himself as a machine or as a robot (or the technical slave). A dispositive stands for a regulation frame, an authority, an entitlement, an opportunity frame, a background structure for actions or an implicit legitimation. It refers to institutional frameworks for actions or to different roles within a collective handling. Technical dispositions, technical roles and so on refer to the horizon-term for their interpretation and to the paradigm-term for the description of different technical practices in the context of the technology concept (Irrgang 2010b).

Technical potentials are the foundation of technical powers. Technical Power is founded on human skills in a context of common technical practice to enable certain technological procedures, which have been more and more delegated to machines since the beginnings of the Industrial Revolution. Potentials include structured fields with a minimum, a maximum and an optimum while potentials are getting realized by their use. Also the competence of regulation has different potentials and varies between minimum and maximum. Technical power is some kind of fluctuation and it depends on the success or failure of technical practice. Technical power is constituted by collective technical handling (technical practice) on the one hand and on the other hand by the authority to dispose over technical artefacts. If these powers do not endure maintenance procedures, technical power is going to come down (Irrgang 2007b). Technical potentials have not been consciously planned in the past. They were something in which human kind finally has sent itself to a development path without knowing the end of it. Heidegger has detected a series of important aspects of the classical techniques and of the modern technologies. But he has hidden those findings behind his word-magic. Heidegger´s philosophical thinking of technology does not replace the philosophy of technology, but it emphasizes the significance of the philosophy of technology.

From a philosophical point of view technology is the motivation for reflection about the technical world into three dimensions. Already Aristotle distinguished in the "Nicomachean Ethics," 1139-1140:

- (1) *Poiesis* as routinized producing practice, which means repeated producing of technical artefacts (executed in the ancient times by slaves, later by labourers). Instrumental knowledge contains supporting service and can be executed by machines as well (assembly line, mass production).
- (2) *Praxis* as technical practice, like its use occurs in planning the house building and executing within constructing the house; creative design oppor-

tunities and individual adjustment (embedding) to the concrete situation; use of technology and embedding in cultural, historical and social contexts.

- (3) *Theoria* within technical practice, at first the machinist and the millwright, the architect and the engineer; Technological and Engineering Sciences, speculation and reflection about technology, philosophy of technology, but also ethics of technology and reflection of an entire technological world in relation to the "lifeworld" (a term that was not used by Aristotle).

Technology of the *poiesis*-type (*instrumental rationality*) can be adopted by machines (spinning and weaving machines, automatic factories, robots). But then it is not a form of technical practice. Different kinds of technologically related theories have to be distinguished from that (Irrgang 2007b; Irrgang 2008a; Irrgang 2009b; Irrgang 2010a; Irrgang 2010b). Aristotle wrote in "Politea," 1st book, chapt. 4:

> "Because [...] single arts and crafts need their own instruments, if their efforts shall succeed [...] Because instruments are partially without soul, partially some with soul, such as for a steersman´s rudder is without soul, but the assistant steersman is an instrument having a soul – because every assistant represents the art and craft of an instrument [...]. Of course, if every instrument could notice its performance after receiving instructions, or even guessing orders in advance, such as the constitutions of Daedalus or the trivets of Hepheaistus have done it [...], then masters would not have further need for assistants and neither lords for servants." (1253b24-1254a8 [Aristoteles 1995, pp. 7-8.])

Within the vision of Aristotle robots and "androids" are replacing the characteristic ancient human technical slave. Slaves and craftsman worked according to instructions (*poiesis* and producing) during ancient time. This kind of handling accords to subordinated labour – the factory labour –, but especially to the labour at the assembly line; to the process of production in the "automatic factory" and earlier to the "automatic weaving machine" which has initiated the Industrial Revolution. These kinds of work are adopted by machines up to smart machines and robots during the technological hypermodern age. But there exists also a sort of technical handling, which is creative. This creative technical handling is important for repairing and maintenance, for modernizing and expanding, as well as for different forms of technical construction. Creative technical handling is subordinated to orders of morality. The one who creates and builds these constructions is responsible for their design. Working is increasingly reduced to the *poiesis* and it is implemented into machines during the Industrial Revolution. That leads to an autonomous technology and to automated factories. As part of the machinery process, labour is reduced to the production process and to the collective instrumental handling, such as working at an assembly line.

The delegation of means and ends to technical systems and automata leads to the instrumental reduction of handling schemata. Robots are higher advanced automata, but they still are automata. There is no fundamental difference between an automatic weaving machine and a humanoid robot. Robots, which are capable of "autonomous" handling, to be more precise which can possibly act similar to an animal in many situations, are still technical products. They remain to be instruments and will not become acting subjects. Handling by robots is a case of handling without any handling subject (Irrgang 2005b; Irrgang 2008a). A depiction of handling schemata within machines has happened since the Industrial Revolution. At that time the first integrated machine transformation of a more complex human-technical handling started with the simulation of the spinning act within machines and the connection of weaving machines. The modelling and simulation of human abilities, including the use of instruments, is therefore nothing fundamentally new. But in fact, something might be possible in the future: The expansion of the basic scheme at non-technical processes, such as perception, mobility etc., which can be used for non-technical purposes. A further technologization of the everyday life will take place and will be enforced by this development. But even this is not something fundamentally new (Irrgang 2008a).

The implementation of operational structures of (human) technical handling into (technical) artefacts leads to an implementation of handling-schemata into machines. Insofar spinning machine and weaving machine have been developed by transforming the process of creation and handling into mechanical production processes. Therefore, technology is based on natural law and verified human procedures and their ability of implementation into technical artefacts (Irrgang 2008a). A formal handling structure separated from the body as behaviour of a robot or as his process procedure is both possible and maybe even programmable. The robot will act within given handling structures. Robots cannot instruct themselves, but are bound to the way of their construction, which intends a given structure of instructions. Their "handling" is pre-interpreted by the way it is programmed. They have no world and do not orientate within a framework of human rationality, but within a frame of given "frames" the same way Artificial Life as technical intelligence is able to do.

Test persons are rating functional products best, but when it comes to paying, real consumer might decide for fancy design and fun. The iPhone could serve as an example for what had to be learned in consumer technology. Consequently, it is no longer sufficient to create an intelligent agent as mechanical partner, which is helping while his counterpart is at a loss and missing out on further knowledge (Technology Review 2/2009, pp. 64-65). Whether in the car, at the hospital or within industrial fabrication – multifunctional, multimodal ways of communication between human beings and machines are catching on. Experts predict a great future for linguistic technology; it is said to already gone beyond a turning point. Multimodality can make life easier to older people. With that it is given an extensive and also social responsibility (Technology Review 2/2009, pp. 70-71). Another role model: Soon, users shall be able to instruct their computers by gesticulation only without using mouse pad, keyboard or touch screen, which means the turn on of instructions by finger gesticulation with customized gesticulation. Admittedly, the disadvantage is that the user would have to learn a new vocabulary for every operation (Technology Review 2/2009, pp. 74-76).

Future machines have to be emotionally designed to interact with humans. Many products are loved by humans because of their behaving designs. Here, it is about a personalization of things. Making something familiar is making a habit out of something (Norman 2004, pp. 215-222). We have to deal with two paradigms if we analyze the use of technical artefacts. One is consumption, which is at the centre of social approaches. The second one is demand, which is focussed at economic analyses. The connection between consumption and demand on the one hand and innovation on the other hand causes a series of interesting questions. It is possible to develop a theory which contains the theory of economic demand based on the cultural-social perspective. Innovations depend on real and on recognized gender embedded processes of creation and production. They facilitate the possibility of innovations (McMeekin et al. 2002, pp. 1-5).

The meaning of social routine for the use of technology can be explained with the basic settings of human beings, while class affiliations are rather defined by capabilities. They can be distinguished into: (1) professionals, (2) managers, (3) routinized "white collar"-labourers, (4) trained "blue collar"-labourers, (5) educated labourers with skills and (6) untrained labour (McMeekin et al. 2002, pp. 79-81). The routinized consume is a describable ability to repeated consumption, which has been learned and acquired by groups of consumers in order to react on social pressure or social contexts (McMeekin et al. 2002, p. 85). Routinized consume and the related changes by new products are more than consumption (sociology) or demand (economy). It is a subject of a philosophical perspective on technology. The usability of a product is therefore equally important to the (cheap) prices of technical products. Shopping is no rational decisive process based on optimized use, which can be calculated in game theory, but it is based on the experience with products in the past, and increasingly on the information from user experiences which are published on the internet. The value a product cannot be judged equally well by all its customers, especially if it was recently released. This handling knowledge goes beyond the biological selection model. Handling knowledge is knowledge about an interaction, which involves the *human-artefact-interface* (Rohracher 2006, pp. 17-23).

The Different Cultural Robot-Traditions in Europe and Japan

The concept of the robot (as automat) belongs to the mechanical tradition of engineers, researchers (example: "Frankenstein") and of the working machine in Europe. The mechanism conceived as something unnatural, with the result that the mechanical automata had often developed a terrible independent existence.

Automata are working machines in Europe. They replace labour although it is a high valued social good in Europe. Changes of our environment by autonomous intelligent technology (automation) could be larger than changes, which are triggered by the humanoid robot which is conceived as a human companion and partner. From a European point of view the more important issues are found in the automation and in the change of the paradigm of work. The robot in Japan, especially the humanoid robot, emerged from the tradition, from folk culture and from pre-scientific myths. As a matter of fact he has not such terrible independent existence like in Europe and is derived from a childlike scheme. Humanoid robots are therefore more popular and socially acceptable in Japan.

During the ancient times they were suggested to be freely deployable and controllable slaves called "androids." *Hephaistos* has forged *Pandora* through godly mission. Even *Daidalos'* artificial created human beings (in the sense of living statues) became traditional since the Greeks. The "Iron Maiden" of the tyrant *Nabis* of Sparta (200 B.C.) was even a real statue. Citizens who did not pay tribute got "hugged" and then speared by her stings if they would not pay on time. During medieval times there have been threatening and inflective statues. The mechanical clock and automatic mechanisms were discovered first at Byzantium and Arabia. The "Iron Man" by Albertus Magnus did appear to serve as his doorkeeper. Vaucansons' flutist entertained the audience in 1738. The 18th-century has been the century of "androids," which occurred within various legends during the age of the mechanics (Völker 1994, pp. 470-471). The flutist was a system of bellows, driven by clock units. *Pygmalion* can be seen as a living statue (Völker 1994, p. 488). Within the Greek mythology the forging goddess *Hephaisto*s was responsible for creating artificial beings. *Daidalos*, the Attic master-builder, belongs to the fabulous mankind creators, as well. His inventions are already technically comprehensible. The term "android" relates to the Greek words "aner" ("man," "human being") and "eidos" ("look," "form," "shape") what stands for "The human copy." Admittedly, he was shaped first during the time of absolutism. During the 17th- and 18th-century the artistic clockmakers promoted the replication of the human being to new heights. But within the Industrial Revolution, the practical interests to create human like machines got lost for the engineers. Mary Shelley's literary character of "Frankenstein" is a counter movement to this. Shelley quotes a basic topic from the literary illustration of technology: the unpredictability of research and invention, with cross natural borders (Drux 2001, pp. 68-71). These discussions are nowadays promoted, facing artificial life and expert systems. Especially the defeat of Kasparow in his match with the computer *deep blue* has caused a sensation. The computer`s game appeared to be intuitively right, inventive and highly intelligent to Kasparow. What a human would have done for his feelings had been calculated through a machine. The world champion concludes that high quantity leads to quality at certain, at least in chess (Irrgang 2005a).

E.T.A. Hoffmann talks about "androids," what means humanlike beings. An artificial human being plays for example the main character in his work "Die Automate" from 1814. The "speaking Turk" is the protagonist, an automat, which is so artistically manufactured that it is impossible to find the source of its voices. The speaking Turk is adequately responding to questions by ingenious and appropriate answers in different languages. Automata are imitations of human beings, which are almost perfect if they are no longer to distinguish from the original. A certain kind of averseness towards the waxworks becomes apparent in the writings of E.T.A. Hoffmann, even if the instruments of the artists are very reasonable. The automaton and the mechanical orchestra of the professor, consisting of male and female dancers, are displaying the dead and numbness of the machines music. The mechanic is using his ability (art) for this adverse joy and not for the perfection of musical instruments even though the perfection of the musical instruments would lead to higher musical mechanisms. Actually, it is about finding the perfect tone. Natural noises and environmental music are put into one context. Consequently, the saying of the Turk turns out to be a lie. Hoffmann´s conclusion is that the human being is not replaceable via technology. The automaton is not the right execution of technology. Hoffmann plays with the ancient and romantic motive: technology is limited. Technology should be used to build musical instruments, and therefore it is decreed as pure device. Technology has to be used and controlled by the human being. But the human being should not be replaced by technology. In E.T.A. Hoffmann´s "Der Sandmann" is an "android" as well, same as in Ambroce Bierce´s "Moxon´s Master" and as in "Meister Zacharias" from "Le Docteur Ox" by Jule Vernes (Irrgang 2005a).

Robots, cyborgs and other "androids" are playing a special role in the science-fiction literature of the past decades. During the era of Reagan emerged a movement of massive anti-technological sentiment. It was caused by the Vietnam War, which just reverted into renewed optimism, while new technologies promised an economical strengthening (upturn). Models have been formulated in Star Wars where the scheme of a robot was transferred into a defence system. This led to a proceeding militarization of the universe. A cyborg contained subjectivity in the "theatre of mind." The movie "The Terminator" has been characteristic for that. Cyberspace is an artificial world within one computer or within a network of computers. It is the matter of an artificial intelligence without human corpus or body. The cyborg represents the abstraction and emotional distance, which is produced by technological media. Computers are also central in the movies "2001 – A Space Odyssey" and "War Games" from 1983. Machines can be obsessed, they can be constructed as servants and act like servants. But there is a counter-movement which assumes subjectivity for machines. In this way the cyborg can get rehabilitated. Such "android" commanders are for example displayed in the movie "Blade Runner," which is about "replicants." There is a police officer who hunts such human clones but until the end it remains uncertain whether he might be a replicant as well. The technology which was thought to be external shifted increasingly towards the inside of the human nature. It becomes more and more difficult to say if we can trust our senses if we want to judge if someone is a real human being. Gibson´s "Neuromancer" is some kind of cyberpunk. The connection of all computers is the cyberspace. If everything becomes optional and re-combinable, such as the human body, human experiences, human culture and the reality itself in the "Neuromancer," then everything can be sold and economised. This will lead to a conversion of the green world or rather of the living world into a closed world – this is the program in "Neuomancer´s" future, where whole artificial ecosystems are created (Irrgang 2005a).

The term "robot" relates to Slavonic meaning "work," "compulsory labour" and "forced labour." It follows the science fiction-play "R.U.R." written by Karel Čapek in 1921. The author goes back to the "Golem-tradition" in a vat, where artificially created humanlike labourers revolt against their creators and other humans, enslaving them in the end. In Germany robots have a totally different reputation and local value in everyday life as in Japan. It will be clarified, why robots have such a good reputation in Japan. The term "robot" is the description for a mechanical companion, which often has a replacing purpose, but which also includes industrial robots and automata (Wißnet 2007, pp. 4-10). The term "mechanism" stands for automata as well as for illusions in Japan. It stands for an illusion of the living. *Karakuri Ningyo* means "mechanical doll" and is first mentioned within a collection of ethnical narrations in the year of 1110. These figures are practical and playful. The first robot instruction is described by *Karakuri Zui*, about 200 years ago. It is the description of four mechanical clocks and automata. But it has to be said that Japanese clockmakers were not able to create metal clockworks. The technology working in the background was seen as something magical, positive and cheerful. The first robots were built in Japan in 1929/30. Especially in Japanese children´s books robots have been shown as friendly, lovely and funny companions. Japan´s rise to be the robot nation started in the 1960s (Wißnet 2007, pp. 19-37).

One important contribution to the development and acceptance of robots comes from the imaginary world which is found in *Mangas* and *Animes* based on *Titsuawan Atomo* (Astro Boy).[1] He influenced the picture of the robot in Japanese Comics. Astro Boy is the figure of a robot which appears to be a five-year-old-child who already has a soul. The development of humanoids has a more positive connotation in Japanese public than anywhere else. They focus on the potential of cost savings through robots. The appointment of guest-workers within labour deficits is completely refused in Japan while the "robotization" of the society is preferred. The entire Japanese community keeps on growing older

1 See also the chapter "Robots in Japanese Popular Culture" (Maika Nakao) in this book.

and this will cause a heavy financial burden for the working part of the society. Robots are utilized within elder care and nursing. "Regina O-II" has been such a robot. The communication-robot "ifbot" first lead to a couple of acceptance issues, because older people are not much open-minded towards innovations. This also comes along with high costs. A robot that helps changing clothes or doing personal hygiene occurs to be uncomfortable as well. An overall ideal kind of a supporting robot does not seem to exist yet, but robots have to be seen as a key technology for the future.

Japan is leading in the construction of humanoid robots. Some people see the Japanese toilet as the preliminary type of robots. Its story began in the 1970s. There have been models constructed which rinsed off the bottoms of the Japanese by pushing a button. The notion of "toilet" almost offends the newest model generation of Japanese producers, such as ToTo, Inax or Panasonic. Through approximation sensors and weight sensors, motors and mechanisms the toilets where transformed from lavatories to almost maintenance-free robo-toilets. For example, the newest model welcomes his user while he opens the toilet lid. The 2750 Euro high-tech toilet evacuates the exhalation and releases them deflavoured into the ambient air during the actual business does. After the bowel movement it performs the regular bottom cleaning, including drying by blow-dryer, flushing and closing the toilet lid. Therefore, it is no wonder that Japan takes a leading role within the development of other service robots. The most interesting project is currently performed by Shigeki Sugano in Tokyo. But it displays the limits of this intent, as well. He designed a prototype of the capacious service robot "Twendy – One." This adventure could have cost the company some ten million dollars. But Sugano admittedly tests the utility opportunities under real circumstances for "Twendy – One" and its up-coming commercialization. Sugano´s first results take Japanese robot dreams right back to the ground. The belief that millions of partner robots will populate flats in the future, or that all-rounders apply to every annoying housework such toilet cleaning, washing dishes, vacuum cleaning or ironing, remain to be a dream for now. Most of the planned every day work does not have to be mastered by sub workers (1) because they should not be programmed into robots and (2) because programming such processes into a service robot would be far too expensive and too complex (Kölling 2009, pp. 74-75).

The Robot as Sensitive Partner

"Affective Computing" shall transform soulless machines into perceptive partners. In the beginning Ishiguro, a Japanese designer and engineer, consequently tried to create his robots as humanlike as possible in order to achieve this empathic effect. A real emotional interaction between user and computer can only emerge if the machine is able to transmit an emotional authenticity without being controlled by a human. But there are about 100 different definitions of the term "emotion" within psychological literature. Other researchers think that this emotional model is completely wrong and they argue that the commitment of basic emotions derives from a cultural background. This is why they prefer the so-called "dimensional emotion model," which is not reducing emotions onto a few standard types, but which is representing conditions (their transitions are fluent) within an abstract space of emotions (Stieler 2010, p. 38). It is not important what is felt by human or machine if we try to differentiate one emotion from another, but it has to be taken into account to whom and what the emotion is related and which consequences it is causing within different plans and propositions of the machine. It is therefore not further remarkable that a robot looks like a human. He has to achieve instead certain standard schemata. It is logically consistent that "Flobi," another model of a humanoid robot, is no longer designed like human, but like stylized comic figure: big, round saucer eyes, small chin, round face – the entire thing looks like an oversized toy. Is this the personalized child scheme? The robot should not appear to be a thread. The exploration of social relations between humans and robots are so far advanced within interdisciplinary researching projects at the University of Bielefeld that they aim to motivate the astronauts with robots for practice during a new cooperation with the German Centre of Aerospace at Cologne (Stieler 2010, 40-41).

But the emotional software, which understands the user and responds to him, is not yet developed. Currently, most humanoid robots look like humans or like children in a spacesuit. The interactions between humans and such humanoid robots are analysed in elaborate research institutes. What the players do not know: every move of them, their position in the room, the distance and reaction to the machine is recorded and evaluated by scientists and engineers (Stieler 2010, p. 42). From a philosophical point of view the "Joy of Use" can be analysed methodologically as the empathy for other humans in the second-person-perspective (2PP). This capability of humans depends on natural dispositions, which are traceable within our brain. They are for example found in so-called "mirror neurons," which enable us to feel the pain of others for ourselves. They allow us to do multiple distinctions and they lead us to a transcendental level of appreciation for other humans or for possible substitutes like a beloved pet. This behaviour is nowadays already known as "pet syndrome." The human-machine-interaction is a technical potential. It can also be described in terms of the "Joy of Use" as the emotional relation, which is constituted in the human-machine-interaction (Irrgang 2010a).

"Affective Computing," "Emotional Machines" and different types of learning-software enable humans to interact with machines. The second-person-perspective (2PP) from phenomenology and neuroscience (mirror neurons) has to be explicated. The goal is to develop a theory of the mental states of the opponent (other minds). The intelligence of humans, animals and artefacts are three different levels or stages of embodied knowledge (Irrgang 2007a; Irrgang

2009a): "Love your robot!" The Philosophy of Technology and the pet syndrome are one side of the coin, to fear technology or to have a phobia in front of certain animals (spiders) is the other side. It is about copying the human being. It may be recommendable to us a figure with a childlike scheme or comic figure like in Japan to care for the elders.

The *hetero-phenomenology* has reached a new dimension in the age of the technological "Uebermensch" and it has become separated into three parts: *hetero-phenomenology* of other bodies (human intelligence), other living intelligences (the intelligence of organisms, mostly animals) and other technical intelligences. This gradualism works like an analogy. It perceives other bodies in their sensitivity to pain, while pain remains to be a simulation in this sense. It is also about the emotions of other bodies, which may be shaped actively. Other bodies in terms of different behavioural patterns are modelled in handling simulations. The traditional phenomenology is different from a phenomenology, which sees the things as emerging. The *hetero-phenomenological* perspective accepts three kinds of other intelligences in the sense of a trans-classical phenomenology.

- (1) The other human being as a person (with consciousness and awareness of his future, his death and himself): *Human Intelligence*;
- (2) Other biological intelligence like the intelligence of animals (with consciousness, possibly self-consciousness): *Natural Intelligence*;
- (3) Autonomous acting machines (without any model of themselves and without consciousness): *Artificial Intelligence* (Irrgang 2009b).

So emotional Software remains to be science fiction and it is quite difficult to be realized. We need to search for existential orientations of humans which are able to express feelings. Socio-Biology has already tried to find such stereotypes in behavioural patterns some 30 years ago, but it was not very successful in describing a framework of psychological drives and instincts. There already has been an emotional relation to technology before. The human-machine-interaction has to be understood as a potential of the human-body and as a technical potential. This interaction allows the simulation of human activity on its lowest level, which is the *poiesis*-structure. This is necessarily some kind of abstract emotionality or some kind of calculated affectivity. It is possible to love technical artefacts just as it is possible to love a pet, but it is a dictum that service robots are not forced upon anybody. The robot is seen as sensitive partner within the development of "Affective Computing." In this sense he is sending non-verbal signals to humans, animals and techniques. The development of a theory about the status of the opponent or about the mental states of conversation partners or cooperation partners is always taking place.

Maybe we cannot teach self-esteem to a robot, but we can embed calculable behavioural patterns as reactions to certain situations: we can embed situational coded tasks, if we learn to calculate and to program the corresponding key stimuli. Even recursive learning of behavioural patterns might be possible. But this is not the same as human learning. Because there is no algorithmic simulation of implicit knowledge, no generation of knowledge and understanding, maybe not even the recognition of causal correlation. There is no knowledge which contains parts of the first-person-perspective (1PP) and the second-person-perspective (2PP). Without that it would not be human knowledge. Body reactions can be detected by robots, but they cannot develop a feeling for their body (Irrgang 2007b; Irrgang 2009a).

Conclusion: The Search for Appropriate Machines to Cooperate with Human Beings

In comparison to the number of installed units per year, industry robots cause more accidents than regular machines like the press which is classified as dangerous. Why should we build universal applicable robots, if there are humans who can accomplish their work? Space is one of those areas, where robots cooperate better with robots, while human life would be at risk. People will think about such application areas, but we need to know the risk-potential of limited working robots within specialized areas.

It is justified to ask, whether robots have to be humanlike? They can be useful simulations of every day humans but the potential of the human is much more complex as the common sense of the mechanized daily routine. It is not sure if robots can achieve a consciousness of their actions or if they will be able to plan and decide like humans, not even if we achieve to program their action-sequences or some sort of learning process (Irrgang 2005a; Irrgang 2008b). We also do not ascribe a calculator that it has the necessary consciousness to control the content of arithmetic operations, just as text processing programs probably do not understand the texts they work with.

We have to ask whether robots are able to act and how their actions can be evaluated, subsequent to a phenomenological understanding of our bodies. There is no doubt that robots can perform certain handling-schemata, if we see their mode of behaviour. A mode of behaviour is given in the most common sense, if an intention for and the goal of an action are given. The structure of an action in terms of a course of action and a goal of that same action has to be programmed for robots, for example with an instruction like "collect items on Mars!" An acting subject is not necessary to accomplish a given goal. The indicators of an action can exist even though there is no acting subject (Irrgang 2005a; Irrgang 2005b). Utilization is based on redundancy. A powerful recognition of our common speech is necessary for an acceptable robot-technology. Problems with data protection and data security are also connected to this use of modern technology. Legal questions will be emerging as well as marketing issues confronting consumers who already have almost everything they need. How can the first users be won? Intelligent domestic appliances and networks are necessary therefore. We will buy a lot of things online instead of going to shops in order to look what else they have to offer. So, new marketing strategies will be necessary as well. The crash of a system of personal computers could also lead to total chaos, in which we couldn`t even call for help by phone. So will we end up in a technological disaster and choke in technology? Probably we will not. The client has to learn to ask what he really needs and the producers have to learn from it and to produce appropriate technology for humans. The dictate of the moment is to create technology that works unobtrusive and that really supports us (Zühlke 2005).

Usability is what Software-ergonomics aim for. The different operating levels have to be distinguishable. We have to distinguish between (1) simple users, (2) experienced users and (3) specialist and mechanics. Hardware development was a dominant field of investigation from the 1950s to the 1970s, in the 1980s a Software-crisis and a scientific shift towards software-problems occurred. There is no understandable structuring, no proven method and no approach to assure quality in this field so that we can almost speak of a crisis of usability nowadays. The developers have never known what usability really means and firm studies of the users are still missing.

Development processes are very cost intensive and a lot of things are not investigated therefore but *Useware* has to focus on the design of operating systems. Innovations and usability of artefacts still focus on the first groups of users. It is also remarkable that aggressive marketing methods are trying to create a sort of hypnosis, which shall make the client think that he has to jump on the bandwagon as fast as possible, if he does not want to be one of the sad persons who are outdated. Innovators and first users love the hype but the "Joy of Use" is related with a simple usability and at least as well with a humane technology (Zühlke 2005).

References

Aristoteles 1995: Politik. Übersetzt von Eugen Rolfes. Aristoteles Philosophische Schriften in 6 Bänden, Hamburg.

Drux, R. 2001: "Künstliche Menschen." In: Spektrum der Wissenschaft 6/2001, pp. 68-75.

Irrgang, B. 2005a: Posthumanes Menschsein? Künstliche Intelligenz, Cyberspace, Roboter, Cyborgs und Designer-Menschen – Anthropologie des künstlichen Menschen im 21. Jahrhundert, Stuttgart.

Irrgang, B. 2005b: "Ethical acts (action) in robotics." In: Brey, Ph., F. Grodzinsky & L. Introna (eds.) 2005: Ethics of New Information Technology. Proceedings of the Sixth International Conference of Computer Ethics: Philosophical Enquiry (CEPE2005). July 17-19. Enschede. The Netherlands, Center for Telematics and Information Technology (CTIT), Enschede, pp. 241-250.

Irrgang, B. 2007a: Gehirn und leiblicher Geist. Phänomenologisch-hermeneutische Philosophie des Geistes, Stuttgart.

Irrgang, B. 2007b: Technik als Macht. Versuche über politische Technologie, Hamburg.

Irrgang, B. 2008a: Philosophie der Technik, Darmstadt.

Irrgang, B. 2008b: "Intersubjectivity, 'other Intelligences' and the philosophical Constitution of the Human-Robotics-Interaction." In: Prajna Vihara, Vol. 9, Nr. 2, July-December, 2008, pp. 56-69.
Irrgang, B. 2009a: Der Leib des Menschen. Grundriss einer phänomenologisch-hermeneutischen Anthropologie, Stuttgart.
Irrgang, B. 2009b: Grundriss der Technikphilosophie. Hermeneutisch-phänomenologische Perspektiven, Würzburg.
Irrgang, B. 2010a: Von der technischen Konstruktion zum technologischen Design. Philosophische Versuche zur Theorie der Ingenieurspraxis, Münster.
Irrgang, B. 2010b: Homo Faber. Arbeit, technische Lebensform und menschlicher Leib, Würzburg.
Kölling, M. 2009: "Vom Klo zum Robobutler." In: Technology Review 9/2009, pp. 74-75.
McMeekin, A., K. Green, M. Tomlinson & V. Walsh 2002: Innovation by Demand: An Interdisciplinary Approach to the Study of Demand and Its Role in Innovation, Manchester & New York.
Norman, D. 2004: Emotional Design. Why we Love (or Hate) Everyday Things, New York.
Rohracher, H. 2006: The mutual shaping of Design and Use: Innovations for sustainable Buildings as a Process of social Learning, Graz.
Stieler, W. 2010: "Mein Freund, der Roboter." In: Technology Review 11/2010, pp. 36-42.
Technology Review 2/2009: "Fokus Benutzerschnittstelle." In: Technology Review 2/2009, pp. 60-76
Technology Review 4/2010: "Fokus Robotik." In: Technology Review 4/2010, pp. 57-75.
Völker, K. 1994 (ed.): Künstliche Menschen. Dichtungen und Dokumente über Golems, Humunculi, lebende Statuen und Androiden, Frankfurt a.M.
Wißnet, A. 2007: Roboter in Japan. Ursachen und Hintergründe eines Phänomens, München.
Zühlke, D. 2005: Der intelligente Versager. Das Mensch-Technik-Dilemma, Darmstadt.

(Translated by Silvio Wende)

Roboethics and the Synthetic Approach – A Perspective on Roboethics from Japanese Robotics Research

Kohji Ishihara

Introduction

The term "roboethics," which refers to ethics as applied to the field of robotics, was coined by Italian robotics researcher Gianmarco Veruggio in 2002 (Veruggio & Operto 2008, p. 1504). This term is a family member of a group of new terms relating to the ethics of technology, such as "neuroethics" and "nanoethics" (Ishihara & Fukushi 2010). Roboethical issues are not new; they have always been a major theme in the literature of science fiction. Indeed, Karel Čapek´s play "R.U.R." ([1920] 2004), in which the term "robot" was coined, dealt with the ethical dilemmas raised by manufacturing humanoid intelligences. Moreover, it is well known that Isaac Asimov´s stories about robots beginning in the 1940s were the source of the Three Laws of Robotics (Asimov [1979] 1991). However, it was not until recently, perhaps beginning in 2004, when the first international symposium on roboethics was held in Italy and a roboethics committee was established at the Institute of Electrical and Electronics Engineers Robotics and Automation Society (IEEE-RAS), that ethical issues concerning robots came to be regarded as topics of academic discussion (Veruggio & Operto 2008, p. 1504). Roboethics has common characteristics with neuroethics and nanoethics: (1) it deals with a new, emerging field where notable applications in the real world are expected, and (2) it attempts to specify and discuss ethical issues before they enter public awareness (Ishihara 2009, pp. 20-21). However, of course, there are some issues that are specific to roboethics.

In the field of robotics, particularly in Japanese robotics, the *synthetic approach* has played an important role. In this paper, I will briefly sketch the history of Japanese robotics and show how it has been related to the *synthetic approach*. Of course, it is beyond the scope of this paper to describe the complete history of Japanese robotics or survey the various views of many Japanese robotics researchers. Instead, I would like to discuss several leading researchers and identify some aspects of Japanese robotics, which will shed light on roboethics from Japanese perspective.

The Synthetic Approach and Ethical Issues in Robotics

In his seminal work, Gianmarco Veruggio (2005) classified six fields of roboethics: (1) "economy" (disruption of employment as human workers are replaced by robots), (2) "effect on society" (including human dependence on and addiction to robots), (3) "healthcare" (bioethical issues in using robots in the field of health care), (4) "lack of access" (fairness concerning distribution of economic benefits from robots), (5) "deliberate abuse/terrorism" (use of robots as weapons), and (6) "law" (legal responsibility for the actions of robots).

Among these fields, those of "effect on society" and "law" can be considered specific to roboethics. In this context, the field of "law" addresses the question of who is responsible for actions taken by autonomous robots. If autonomous, embodied robots are to interact with humans in the real world, new problems of safety and responsibility will arise. Who should be responsible if autonomous robots harm people or goods? This problem is closely related to that of the moral status of robots. If robots with sufficient intelligence and autonomy are created, people might feel they should be granted a special moral status, perhaps not the same as that of humans but higher than that of machines or animals. If this is the case, the concepts of responsibility and autonomy among humans and robots will become very complicated.

In order to avoid this problem, a Japanese robotics researcher, Susumu Tachi, proposed three components of robotics from the perspective of roboethics (he uses the term "robot ethics"): "safety intelligence," "non-anonymity," and "the idea of robots as shadows" (in Japanese, "bunshin") (Tachi 2002, pp. 199-203). Safety intelligence is the specific intelligence that is necessary for workplace safety in cases where robots and humans work together. Safety intelligence recognizes where humans are, predicts the movement of robots if they obey human operation, and rejects operations that will be dangerous to humans (Tachi 2002, p. 181). Non-anonymity refers to the clear identification of the human users who operate robots, which manifests human control over robots. This concept is related to the last concept, "shadow," which refers to the enslavement of robots to humans. Tachi contrasts the concept of "shadowiness" with that of "robots as Others," which is used to indicate autonomous robots with will and emotion. Tachi suggests that relegating robots to the role of Other is problematic not only from the perspective of responsibility but also from the ethical perspective.

The problem of "robots as Others" vs. "robots as shadows" (henceforth, the latter will be called "robots as servants") is related to Veruggio´s field of "effect on society." Veruggio characterizes these issues in the following manner:

> "What is it going to happen when these smart robots will be our servants and house stewards, and when our lives will depend on them? Technology addiction to robots can be more dangerous than to TV, Internet, and videogames" (Veruggio 2005, p. 4).

This field addresses the possibility of embarrassment or fear that derives from the presence of intelligent, autonomous robots in society, who are comparable in sentience to humans. Here, the embarrassment and fear come from the possibility that robots might threaten human superiority and governance, and also that some humans may become so dependent on or addicted to robots that they avoid or dispense with human interaction. In order to avoid these problems, Tachi suggests that we should pursue the path of "robots as servants," that is, robots as tools and enhancers that are fully under human control.

As another aspect of "effect on society," we could consider the concept of "robots as mirrors." Robots have already reflected and influenced human understanding of humans and humanity. A well-known Japanese robotics researcher, Hiroshi Ishiguro, indicates that the attractiveness of robots arises from the fact that robots are "mirrors of humans" (Ishiguro 2007, p. 17) in the sense that "building robots is understanding humans" (Ishiguro 2007, p. 222). The mirror-character of robots can arouse fear in humans, for humans have the tendency to be afraid of doubled selves, as illustrated by the traditional fear of the "doppelganger" (Anderson [2011] calls this effect the fear of the "uncanny double"). Can robotics researchers ethically build robots that make humans uneasy?[1] How will robotics affect human understanding of humans? These issues must be considered as a part of the field of "effect on society" in roboethics.

The attempt to understand humans by building robots has been called the "*synthetic approach*" (Pfeifer & Scheier [1999] 2001, p. 22). Japanese researchers including Ishiguro (Asada et al. 2001) have proposed the concept of "cognitive-developmental robotics," which adopts the *synthetic approach* but focuses on human cognitive development. Another Japanese researcher proposed "computational neuroscience" which aims to "so deeply and essentially understand functions of the brain as to be able to build computer programs or artificial machines, which are able to reproduce those functions in the same way as the brain" (Kawato 1996, p. 10). In other words, it aims to "know the brain by creating brains" (Kawato 2005, p. 99).

Although it is not until recently that some Japanese robotic researchers have explicitly recognized that their approaches are examples of the *synthetic approach*, the approaches of Japanese humanoid-robot researchers have been synthetic from the very beginning. It is remarkable that at an early stage of research and development, Japanese robotics researchers aimed to create bipedal, walking humanoid robots. Walking robots were not developed only in Japan. For example, Ralph Mosher and others at General Electric (GE) developed a "Quadruped Walking Machine" (QWM) in the 1960s (Mosher 1970; Katô 1975), Robert McGhee and others at the University of Southern California developed another quadrupedal walking machine in the 1960s, and D. C. Witt at the University of Oxford developed a bipedal walking machine between 1968 and 1970 (Vukobratovic 1975; Katô 1975). However, although researchers in the US and the UK tended to develop robots in order to extend specific human functions, Katô Ichirô's laboratory at Waseda University in Japan aimed to develop fully humanoid robots from the very beginning. By 1973, they had developed WABOT-1, which was "the first full-scale anthropomorphic robot developed in the world" (Koganezawa et al. 1991, p. 5).

1 Coeckelbergh (2009) discusses "the ethical significance of appearance" of robots, that is, what they look like.

The question is why humanoid robots have been developed mainly in Japan until recently (until 1993, when Brooks at MIT began to build "Cog." See below). It is difficult to answer this question clearly, although we can identify many possible factors: religious background, cultural background, structure of Japanese industry after World War II, development of industrial robots, research focuses and environments of research laboratories at Japanese universities, influence of manga and anime, etc. However, it is beyond the scope of this paper to scrutinize these factors. Instead, I will consider some aspects in order to elucidate the implications of the *synthetic approach* and suggest its significance to roboethics.

Karakuri and Biomechanisms

The Japanese mechanical doll (*Karakuri*) tradition is often regarded as a conceptual background of modern Japanese robotics.[2] *Karakuri* were first produced in 1662, when Takeda Ômi, a clockmaker, built the Takeda Karakuri-za (*Karakuri* house) in Osaka (Tachikawa 1969, p. 162). *Karakuri* developed further over the next 100 years, and an excellent technical book, *Karakuri-zui*, was published in 1796. European counterparts of *Karakuri* are the automata made by Jacques de Vaucanson and those made by Pierre Jaquet-Droz in eighteenth century (Katô 1982: 40). Although both Japanese *Karakuri* and European automata were made for entertainment, their traits are very different: *Karakuri* dolls were abstract and stylized while European automata were realistic (Katô 1982, p. 42).[3]

In 1967, a wind-up tea-carrying doll illustrated in *Karakuri-zui* was reproduced by Shôji Tachikawa at Kitazawa University and students at Waseda University (Tachikawa 1969, pp. 223-228; Schodt 1988, pp. 60-62). The reproduced doll was acclaimed and Tachikawa´s short introduction of *Karakuri* (Tachikawa

2 Pauer (2010) gives a detailed history of *Karakuri* and points out that there is no direct connection between modern robots and *Karakuri*. Pauer´s paper is included in "Technikgeschichte´s" special issue on the cultural history of robots in Japan. In the introduction of this issue da Rosa and Pauer (2010) give a concise history of Robots in Japan. Matsuzaki (2010) tries to explain the Japanese attitude toward humanoid robots referring to the "egalitarian traditions of Buddhist and Shinto thought" and Itô (2010) discusses the background of the images of robots in Japanese manga after World War II.

3 Katô argues that under the influence of Confucianism, technical expertise was not valued in the Edo period (1603-1868). Suzuki (1988) indicates that government policy influenced the development of *Karakuri*. In 1721, the Tokugawa government stipulated the law Shinki Gohatto ("Ban on Novelty"), which prohibited the introduction of new things. This law was established in order to promote frugality, and outlawed technological innovation. However, show business, including *Karakuri* shows, was an exception. As a result, *Karakuri* developed for entertainment, isolated from industrial applications (Suzuki 1988, pp. 41-42; Umetani 2005, p. 35). Yôji Umetani, a Japanese robotics researcher, indicates that most contemporary Japanese robotics researchers have been similar to *Karakuri* masters in that they followed their individual academic curiosity without regard for industrial needs (Umetani 2005, pp. 36-37).

1972) with figures extracted from *Karakuri-zui* and photos of existent as well as reproduced *Karakuri* dolls, opened the first issue of the journal of the Society of Biomechanisms Japan. This society was established in 1968, 15 years before the foundation of the Robotics Society of Japan and led Japanese robotics research at the initial stage. The original name of the society was "Jinkô no te kenkyûkai" (the "Society of Artificial Hands"), because it was established as an outgrowth of a research project titled "Jinkô no te ni kansuru kenkyû" (research into artificial hands) proposed by Masahiro Mori and Ichirô Katô (Katô 1987; 1994).

Biomechanisms and the Uncanny Valley

Biomechanisms is the discipline that makes use of biological knowledge for engineering (Katô 1972, p. 1). According to Katô,

> "[biomechanisms] aims to develop healthcare engineering, technology for everyday life, and manufacturing technology in the fields of cyborg engineering and robot engineering; it makes much not only of analysis but also of synthesis" (Katô 1994, p. 5).

It is noteworthy that Katô indicates here the importance of synthesis. We may say that the approach of biomechanisms research was an example of the *synthetic approach*. Indeed, as early as 1968, Masahiro Mori, another leading figure of the Society of Biomechanisms Japan, emphasized the importance of synthesis (making):

> "[A]nalysis of humans is not sufficient to understand humans. Making humans is the way to understand humans [...]. This is true not only with regard to intelligence but also to an eye or a finger" (Mori 1968, p. 872).

However, Mori's interest was not directed at humanoid robots, unlike that of Katô. Mori states:

> "I distinguish between robots as buddies (*nakama*) and robots as slaves. The former are based on the human instinct to make their buddies by technological methods [...] and descend from the tradition of marionettes. The appearance of this kind of robot should be similar to that of humans, for this kind of robot should raise human affinity. Katô at Waseda University calls them 'natural robots.' These robots, which appear in manga or exhibitions, are not our goal." (Mori 1968, p. 872)

In his short essay that was published in 1970, Mori discussed the problem of affinity in robotics and coined the term "Uncanny Valley," which has become very popular among robotics researchers. Mori (1970) indicates that if the similarity of a robot to humans is low, affinity increases as similarity does. However, if similarity reaches a certain level, the affinity is lost suddenly, falling to the bottom of an "Uncanny Valley." Mori mentions various examples: industrial robots, humanoid robots, dead people, electric arms, *Bunraku* puppets, and ill people. The most uncanny thing is the dead; the electric arm is uncanny because it is very (but ultimately insufficiently) similar to living human hands. We may avoid falling into the "Uncanny Valley" by making electric hands either sufficiently similar or deliberately dissimilar to living hands. Mori suggests that the latter path should be followed:

> "The author recommends [that we look to] the summit on the left side of the uncanny valley, where remarkable affinity through moderate similarity can be realized, rather than taking the risk [of falling into the bottom of the uncanny valley] by pursuing more similarity. In fact, I acknowledge that the possibility of comfortable affinity can be realized on other dimensions than that of similarity, that is, by 'inhuman design,' deliberately not imitating humans." (Mori 1970, p. 35)

Artificial Intelligence and Cybernetics

It is remarkable that in the same year as Mori's paper, Ralph Mosher at General Electric published a paper entitled "Robots that are Extensions of Man" (1970). Mosher refers to Norbert Wiener's thesis, The Human Use of Human Beings ([1950] 1954) as demonstrating the advantage of cybernetic mechanisms. Human use of human beings is "to have [humans] do what they are uniquely qualified to do: supply intelligence, judgment, [and] decision-making capabilities" (Mosher 1970, p. 361). In Mosher's view, the aim of the development of robots is to extend human capability by replacing humans with robots that can do inhuman labor, and not to reproduce human abilities.[4]

Of course, there were efforts to reproduce human abilities in the US. Since the late 1950s, US researchers have been working to develop artificial intelligence (AI) (Russel & Norvig 1995, pp. 16-17). However, the approach of AI research in the early period was disembodied.

English mathematician Allan Turing (1950) proposed what later came to be called the "Turing test." Turing worked out an imitation game in which the interrogator must determine whether his or her communication partner (via teleprinter) is a human or a computer. In this game, participants are prevented from seeing, touching, or hearing the partner. This arrangement "has the advantage of drawing a fairly sharp line between the physical and the intellectual capacities of a man" (Turing 1950, p. 434). This line between physical and intellectual capacity should be drawn not only because "[n]o engineer or chemist claims to be able to produce a material that is indistinguishable from the human skin" but also because "even supposing this invention available we should feel there was lit-

---

4 Cybernetic artificial bodies have been developed in more sophisticated ways than Mosher foresaw. In Japan, the Sankai laboratory at the University of Tsukuba developed Hybrid Assistive Limb (HAL) on the basis of cybernics. Cybernics is a new interdisciplinary research domain advocated by Yoshiyuki Sankai, which is "centered on cybernetics, mechatronics, and informatics, and integrates neuroscience, robotics, systems engineering, information technology, 'kansei' engineering, ergonomics, physiology, social science, law, ethics, management, economics, etc." (Sankai 2010, p. 1). Although the achievement of the HAL project is brilliant, its basic philosophy is not much different from that of Mosher's CAMS project. The aim of HAL is to "expand, augment and support [the] physical capability" (Ibid.) of humans.

tle point in trying to make a 'thinking machine' more human by dressing it up in such artificial flesh" (Ibid.). Thus, Turing assumes that (1) the physical and intellectual capacities of a person can be distinguished from each other, and (2) the external features of a machine should not influence judgment of the machine′s seeming humanity. Such suppositions were shared by US researchers in the early decades of AI research.

Although Turing promoted the development of thinking machines without bodies, Mosher aimed to develop artificial cybernetic bodies to extend the human capacity of machines, excluding the intelligence part. Both approaches avoid reproducing human beings completely – with both body and intelligence – but in opposite directions. Turing felt that the body is not essential if the goal is reproducing human intelligence. Mosher advocated developing robots to extend human beings′ capacity in physical tasks and not to create intelligent robots that are capable of replacing humans.

In AI research, there has been a shift in methodology. Although AI research in the early period aimed to reproduce human intelligence without the body using symbol manipulation, in 1980, Rodney Brooks at the MIT AI laboratory began to adopt new approaches that emphasized the importance of situatedness and embodiment for AI. Brooks′ idea was to produce "architectures for intelligence that are networks of simple computational elements," without resorting to "the traditional uses of central, abstractly manipulable or symbolic representations" (Brooks 1999). Brooks made insect robots that could walk and interact with the environment autonomously using this approach, which he called "subsumption architecture." This approach is also called "developmental robotics," in the sense of a development from "insect-level intelligence" to eventually reaching high-level intelligence (Pfeifer & Bongard 2007, pp. 44-45). After making mobile robots, including the insect robot Genghis, and some column-shaped robots in the 1980s (Brooks 1999, pp. 119-127), Brooks and colleagues began to build their first humanoid robot, Cog, in 1993 (Brooks & Stein 1994; Brooks [2002] 2003, pp. 68-69). Cog is an upper-torso humanoid with 21 degrees of free movement and various sensory systems (Brooks et al. 1998). Brooks and colleagues indicate that classical AI research held false assumptions relating to the study of human intelligence: "reliance on monolithic internal models, on monolithic control, and on general purpose processing" (Brooks et al. 1998, p. 54); they also indicated that the critical attributes of human intelligence are "developmental organization, social interaction, embodiment and physical coupling, and multimodal integration" (Brooks et al. 1998, p. 53). Cog was built to serve as a platform for examining this concept of human intelligence. Using Cog and other robots, Brooks and colleagues conducted many studies on the implementation of joint attention, theory of mind, and other factors in social interactions (for example, see Scassellati 2000; 2001; Fitzpatrick et al. 2003).

Brooks notes that "[w]hen we stated the Cog project, there were hardly any robots with humanoid form outside of science fiction" and "[t]he one exception was at Waseda University in Japan" (Brooks [2002] 2003, p. 69). As mentioned above, the Katô laboratory began to build humanoid robots in the early 1970s. Pfeifer and Bongard indicate that

> "[w]hile in Japan humanoid robots had been a research topic for many years already [before the Cog project by Brooks and colleagues], these activities were not directly related to artificial intelligence. This seems to be the reason why Brooks' move into humanoids had a strong impact on the research community [...]." (Pfeifer & Bongard 2007, p. 45).

However, it would be incorrect if one says that Japanese researchers in humanoid robotics had no interest in AI. Katô called WABOT, the first humanoid robot developed by the lab in the early 1970s, an "anthropomorphic intelligent robot" and claimed that it had intelligence corresponding to that of roughly a one-and-a-half-year-old child (Koganezawa et al. 1991, pp. 4-5). The goal was to develop a humanoid robot intelligent enough to communicate with humans and serve in the service industry as a "personal robot."

Uncanny Valley and Mirror of Humanity

Humanoid-robotics research has continued in Japan through the 1970s, 1980s, and 1990s. For example, Waseda University developed the musician robot WABOT-2 in 1984 and the new bipedal walking robot WABIAN in 1997 (Humanoid Research Institute, Waseda University 2005).[5] In December 1996, Honda Motor Co. suddenly announced "the world's first self-regulating, two-legged humanoid robot, called P2" (Hirose & Ogawa 2007, p. 13). Honda had begun humanoid-robotics research in 1986, but had not previously announced it. Honda's disclosure of P2 had a great impact on society and robotics researchers, in two ways. First, the robot's technical performance was surprising good. It walked very smoothly as compared with previous humanoid robots; moreover, it walked autonomously, that is, without cable connection.[6] Second, the fact that a private company had developed humanoid robots was surprising. It gave the impression that humanoid robots could be industrialized. In 2000, Honda announced ASIMO, the successor of P2, and Sony announced Qrio, following their famous pet robot AIBO (although Sony withdrew from R&D of robots in 2006); other companies such as Hitachi, Fujitsu, Toyota, and Kawada Industries have

---

5 Researchers at Waseda University are now conducting the project "Global Robot Academia," which aims at the "realization of cohabitation between humans and robots." (http://www.rt-gcoe.waseda.ac.jp/members/greeting.html [accessed September 20, 2011]).

6 See a video of P2 walking on the Honda website at: http://www.honda.co.jp/ASIMO/history/honda/index.htm (accessed July 4, 2011).

also engaged in humanoid-robot development (Hornyak 2006, Chapter 7; Yin et al. 2010; Kanda et al. 2011; Yamamoto et al. 2011; Kanehira 2011).

All humanoid robots developed by major Japanese companies seem to have followed the spirit of Mori´s proposal and did not try to bridge the "Uncanny Valley." The shape and movements of these robots are humanlike, but their faces are very robotlike and easily recognizable as those of robots. However, there has been another trend in humanoid robots in Japan. The Kokoro Company, established in 1984 as a maker of robots (especially dinosaur robots) for museum exhibitions, began making humanoid robots for exhibition in 1987. They made many humanoid robots to represent and display the lives of ordinary people or celebrities (Kokoro News No. 58).

Hiroshi Ishiguro (presently at Osaka University) has been collaborating with the Kokoro Company from 2001 to produce many surprising humanoid robots, including Repliee R1, Repliee Q1, Geminoid HI-1, and Geminoid F, which were produced by molding Ishiguro´s daughter, a famous broadcaster, himself, and anonymous woman respectively (Ishiguro 2007; 2009; 2011). The "androids" made by Ishiguro were unique in their realistic re-creation of the appearance and tactile texture of humans.

The direction of development of realistic humanoid robots pursued by Ishiguro seems deviant from the direction of development recommended by Mori. However, according to Ishiguro, the most significant implication of Mori´s hypothesis of the "Uncanny Valley" is that

> "it took the uncanny valley created by robots as a mysterious phenomenon arising from the human function of understanding humans, and [thus] motivated much research in neuroscience and cognitive sciences." (Ishiguro 2007, p. 219)

Indeed, Mori did not fail to mention this aspect of robotics:

> "Why do we have such an uncanny feeling? What is the inevitable reason that such a feeling was endowed in humans? [...] Maybe we ought to start a precise survey to make a map of the uncanny valley in order to understand humans through robotics research and to create human affinity by designs on the inhuman axis." (Mori 1970, p. 35).

However, as shown in this citation, Mori´s idea was that we should investigate the uncanny feeling in order to create human affinity by the inhuman design.

As I have discussed, the *synthetic approach* has been the dominant approach in Japanese humanoid-robotics research since the beginning. However, the term "*synthetic approach*" has been used by Ishiguro in a specific manner to include the "reproduction of humanness" (ningen-rashisa, "similarity to humans") (Ishiguro 2009, p. 66). The first generation of Japanese humanoid-robotics researchers did not pursue the reproduction of humanness. The biomechanisms approach is synthetic; however, it aims to understand biological mechanisms in order to reproduce their functions and build robots as partners or servants, not humanlike artifacts. However, Ishiguro aims to build humanlike artifacts. His deepest question is, "What are humans?" and he builds humanlike artifacts in order to understand humanity.

Interestingly, Ishiguro insists that "[r]obots can reproduce the human mind" (2009, p. 154).[7] Ishiguro also believes that if we breakdown and analyze the minds of these robots, we will find only machines and computer programs. His idea is that in these robots, there is no "evident" mind, that we only project a mind into them, and that human beings in fact also have no evident mind. His stance can be regarded as a kind of reductionism[8] or strong AI in Searle's term (Searle 1980, p. 417). However, at the same time, Ishiguro says that the human mind is uniquely human:

> "A human who doesn't believe in or seek the mind will be a machine [...]. Whatever its essence, the mind is very important for humans. Although I don't recognize the being of the mind, I feel the being of the mind." (Ishiguro 2009, p. 159)

Conclusion

In this paper, I attempted to show that Japanese robotics have adopted the *synthetic approach* from the very beginning. If the *synthetic approach* is regarded as common to engineering in general, it could be considered redundant to say that robotics, as a field of engineering, adopts the *synthetic approach*. However, the *synthetic approach* in robotics raises special ethical and philosophical concerns. As the *synthetic approach* aims at "understanding by building," the question regarding what should be built and reproduced arises; addressing this "what" should be the aim of robotics. If robotics researchers want to reproduce and extend a part of human capability, they should aim to build robots as servants or enhancers under human control. If they want to reproduce parts of human capability related to human interaction, they will want to build robots as "buddies" that communicate with and help humans in the real world. Finally, if researchers want to reproduce humanity as a whole in order to investigate what is human, they will want to build "androids" that cannot be distinguished from humans. As shown above, Japanese robotic researchers have not been unanimous in terms of what directions should be pursued in humanoid robotics. The differences among robotics researchers (as well as between robotics researchers and the public) with regard to which directions should be taken in robotics can raise ethical issues. Roboethics should take into account the discussion of future visions of a desirable society and human relations realized by the introduction of (humanoid) robots.

---

7 Asada also states: "I believe that someday robots with kokoro (mind) – even though it may be not the same as a human mind – with which robots will communicate their minds with humans, will come into the world and alter the image of robots in the world" (Asada 2010, p. 161).

8 In a previous paper (Ishihara 2007), I indicated that the synthetic approach can be roughly regarded as a kind of reductionism.

The central issue with regard to the future human image of robots is the possibility of reproduction of the mind. This issue can be divided into an ethical question and a philosophical one. From the ethical perspective, the question is whether it is ethical for robots to have minds. As suggested above, some Japanese robotics researchers would prefer not to implement minds into robots even if this were to become possible. From the philosophical perspective, the question is whether robots *can* have minds.[9] As Ishiguro (2007, p. 293) suggests, robots with abilities comparable to those of humans are not likely to appear in the near future; therefore, the ethical issues that may accompany their appearance are not urgent. However, the philosophical consideration that has been given to this problem may influence the direction of research and development in robotics; therefore, this problem should be considered a serious social and ethical question.

References

Note: quotes from Japanese texts were translated by the author, Kohji Ishihara.

Anderson, M. 2011: "Through the mirror." (Paper read at the University of Tokyo Center for Philosophy (UTCP), August 1, 2011.)

Asada, M., K. F. MacDormand, H. Ishiguro & Y. Kuniyoshi 2001: "Cognitive developmental robotics as a new paradigm for the design of humanoid robots." In: Robotics and Autonomous Systems, 37, pp. 185-193.

Asada, M. 2010: Robotto to yû Shisô. Nô to Chinô no Nazo ni idomu (Robot as Thought: Challenging the enigma of the brain and intelligence), Tokyo. (Japanese)

Asimov, I. [1979] 1991: "The laws of robotics." In: Robot Visions, New York, pp. 423-425.

Brooks, R. A. & L. A. Stein 1994: "Building brains for bodies." In: Autonomous Robots, 1, pp. 7-25.

Brooks, R. A., C. Breazeal, M. Marjanovic, B. Scassellati & M. Williamson 1998: "The Cog project: Building a humanoid robot." In: Nehaniv, C. (ed.) 1998: Computation for Metaphors, Analogy and Agents, Vol. 1562 of Springer Lecture Notes in Artificial Intelligence, Springer Verlag. (Available at: http://www.ai.mit.edu/projects/humanoid-robotics-group/cog/publications.html [accessed August 13, 2011].)

Brooks, R. A. 1999: Cambrian Intelligence: The Early History of the New AI, Cambridge (MA).

Brooks, R. A. [2002] 2003: Flesh and Machines: How Robots Will Change Us, New York.

Čapek, K. [1920] 2004: R. U. R. (Rossum´s Universal Robots). C. Novack, trans., I. Klima, intro., New York.

Coeckelbergh, M. 2009: "Personal Robots, Appearance, and Human Good: A Methodological Reflection on Roboethics." In: International Journal of Social Robotics, 1(3), pp. 217-221.

da Rosa, C. C. & E. Pauer 2010: "Einleitung (Introduction)." In: Technikgeschichte, 77(4), pp. 295-319. (German with English abstract)

9 My view is that social and ethical strictures will prevent the development of robots with minds indistinguishable from the human mind; that is, the ethical perspective gives the answer to the philosophical question (Ishihara 2008, p. 189).

Fitzpatrick, P., G. Metta, L. Natale, S. Rao & G. Sandini 2003: "Learning about objects through action-initial steps towards artificial cognition." In: IEEE International Conference on Robotics and Automation (ICRA), Taipei, May 12-17, 2003. (Available at: http://www.ai.mit.edu/projects/humanoid-robotics-group/cog/publications.html [accessed August 13, 2011].)

Hirose, M. & K. Ogawa 2007: "Honda humanoid-robot development." In: Philosophical Transactions of the Royal Society A, 365, pp. 11-19.

Hornyak, T. N. 2006: Loving the Machine: The Art and Science of Japanese Robots, Tokyo.

Humanoid Robotics Institute, Waseda University 2005: Untitled (pamphlet), Tokyo. (Japanese and English)

Ishihara, K. 2007: "Reductionism in the Synthetic Approach of Cognitive Science and Phenomenology: Rethinking Dreyfus′ Critique of AI." In: Cheung Chan-Fai & Yu Chung-Chi (eds.) 2007: Phenomenology 2005. Vol. I. Selected Essays from Asia, Bucharest, pp. 211-228.

Ishihara, K. 2008: "Kokoro, nô, kikai. Nokagakugijutsu no genzai (Mind, brain, and machine. Issues concerning current neurotechnologies)." In: Murata, J. (ed.) 2008: Kokoro/Nô (Mind/Brain), Tokyo, pp. 175-194. (Japanese)

Ishihara, K. 2009: "Kagakugijutsu-rinrigaku no tenbô (Outlook on the ethics of technology)." In: Ishihara, K. & T. Kono (eds.) 2009: Kagakugijutsu-rinrigaku no Tenkai (Development of the Ethics of Technology), Tokyo, pp. 9-27. (Japanese)

Ishihara, K. & T. Fukushi 2010: "Introduction: Emerging of roboethics as a new field of ethics of technology." In: Accountability in Research, 17(6), pp. 273-277.

Ishiguro, H. 2007: Android Science: Robotics Research for Human Understanding, Tokyo. (Japanese)

Ishiguro, H. 2009: Robotto towa nanika. Hito no Kokoro wo utsusu Kagami (What is a Robot? Robots as Mirror of the Human Mind), Tokyo. (Japanese)

Ishiguro H. 2011: Dô-sureba "Hito" wo tsukureruka (How Can We Create a Human Android?), Tokyo. (Japanese)

Itô, K. 2010: "Vor Astro Boy. Roboterbilder im Nachkriegsjapan, 1945-1952 (Before Astro Boy: Imagery of Robots in Occupied Japan)." In: Technikgeschichte, 77(4), pp. 353-372. (German with English abstract)

Kanda, S., Y. Yasukawa, S. Nagashima, Y. Murase, I. Watanabe, K. Okabayashi, Y. Hada & T. Morita 2011: "Fujitsu no robot kaihatsu (R&D of robots at Fujitsu)." In: Proceedings of the 29th Annual Conference of the Robotics Society of Japan, 2F1-2. (Japanese)

Kanehira, T. 2011: "Kawada Kogyô ni okeru robottokaihatsu (Development of robots at Kawada Industries, Inc.)." In: Proceedings of the 29th Annual Conference of the Robotics Society of Japan, 2F2-4. (Japanese)

Katô, I. 1972: "Biomechanisms to sono shûhen (Biomechanisms and its background)." In: Biomechanisms, 1, pp. 1-3. (Japanese)

Katô, I. 1975: "Hokô-kikai (Walking machines)." In: Journal of the Japan Society of Mechanical Engineers, 7 (676), pp. 78-84. (Japanese)

Katô, I. 1982: "Karakuri ningyô no nagare (A History of Karakuri Dolls)." In: Journal of the Japan Society of Mechanical Engineers, 85 (766), pp. 40-43. (Japanese)

Katô, I. 1987: "SOBIM no ayumi (Development of SOBIM: Society of Biomechanisms Japan)." In: Journal of the Society of Biomechanisms Japan, 11(4), pp. 137-138. (Japanese)

Katô, I. 1994: "Biomechanisms – past and present." In: Journal of the Society of Mechanical Engineers, 97 (902), pp. 4-7. (Japanese)

Kawato, M. 1996: Nô no Keisanriron (Computational Theory of the Brain), Tokyo. (Japanese)

Kawato, M. 2005: "Nô to ningengata robotto wo tsukuru koto niyori nô wo shiru (Knowing the brain by creating brains and humanoid robots)." In: Doi, T., M. Fujita & H. Shimomura (eds.) 2005: Intelligent Dynamics. 1, Tokyo, pp. 95-180. (Japanese)

Koganezawa, K., A. Takanishi & Sh. Sugano (eds.) 1991: Development of Waseda Robot – The Study of Biomechanisms at Kato [Katô] Laboratory, Tokyo. (Japanese and English)

Kokoro News No. 58, 2001.9: http://www.kokoro-dreams.co.jp/news/kokoroN058.pdf (accessed July 4, 2011)

Matsuzaki, H. 2010: "Gehorsamer Diener oder gleichberechtigter Partner? Überlegungen zum gesellschaftlichen Status von humanoiden Robotern in Japan (Obedient Servants or Equal Partners? On Social States of Humanoid Robots in Japan)." In: Technikgeschichte, 77(4), pp. 373-390. (German with English abstract)

Mori, M. 1968: "Robotto no kihon-shisô to kôsô (Basic ideas and concept of robots)." In: Keisoku to Seigyo, 7 (21), pp. 871-880. (Japanese)

Mori, M. 1970: "Bukimi no tani (Uncanny Valley)." In: Energy, 7, pp. 33-35. (Japanese)

Mosher, R. S. 1970: "Robots that are extensions of man." In: Impact of Science on Society, 20(4), pp. 349-361.

Pauer, E. 2010: "Japanische Automaten (karakuri ningyô). Vorläufer der modernen Roboter? (Japanese automata (karakuri ningyô): Ancestors of Modern Robots?)." In: Technikgeschichte, 77(4), pp. 321-351. (German with English abstract)

Pfeifer, R. & Ch. Scheier [1999] 2001: Understanding Intelligence, Cambridge (MA).

Pfeifer, R. & J. Bongard 2007: How the Body Shapes the Way we Think. A New View of Intelligence, Cambridge (MA).

Russel, S. J. & P. Norvig 1995: Artificial Intelligence. A Modern Approach, Upper Saddle River (NJ).

Sankai, Y. 2010: "HAL: Hybrid assistive limb based on cybernics." In: Robotics Research, The 13th International Symposium ISRR, pp. 25-34. (Available at: http://sanlab.kz.tsukuba.ac.jp/?page_id=61 [accessed August 8, 2011])

Scassellati, B. 2000: "Theory of mind for a humanoid robot." In: First IEEE/RSJ International Conference on Humanoid Robotics, September 2000. (Available at: http://www.ai.mit.edu/projects/humanoid-robotics-group/cog/publications.html [accessed August 12, 2011])

Scassellati, B. 2001: "Investigating Models of Social Development using a Humanoid Robot." In: Webb, B. & Consi T. (eds.) 2001: Biorobotics, Cambridge (MA). (Available at: http://www.ai.mit.edu/projects/humanoid-robotics-group/cog/publications.html [accessed August 12, 2011])

Schodt, F. L. 1988: Inside the Robot Kingdom, Tokyo.

Searle, J. 1980: "Minds, brains, and programs." In: The Behavioral and Brain Sciences, 3, pp. 417-457.

Suzuki, K. 1988: "A positive study on the technique of mechanism in the Edo era." In: Bulletin of the National Science Museum, Tokyo, Ser. E, 11, December 22, 1988. (Japanese)

Tachi, S. 2002: Robotto Nyûmon (Introduction to Robots), Tokyo. (Japanese)

Tachikawa, S. 1969: Karakuri, Tokyo. (Japanese)

Tachikawa, S. 1972: "Karakuri no Sekai (The World of Karakuri)." In: Biomechanisms, 1, pp. A1-A4.

Turing, A. M. 1950: "Computing machinery and intelligence." In: Mind, 59 (236), pp. 433-459.
Umetani, Y. 2005: Robot Kenkyûsha wa Gendai no Karakurishi ka (Are Robotics Researchers Modern Karakuri-Masters?), Tokyo. (Japanese)
Veruggio, G. 2005: The birth of roboethics, ICRA 2005, IEEE International Conference on Robotics and Automation Workshop on Roboethics, Barcelona, April 18, 2005, 4p. (Available at: http://www.roboethics.org/icra2005/veruggio.pdf [accessed February 5, 2010])
Veruggio, G. & F. Operto 2008: "Roboethics: Social and Ethical Implications of Robotics." In: Siciliano, B. & O. Khatib (eds.) 2008: Springer Handbook of Robotics, Berlin, pp. 1499-1524.
Vukobratovic, M. 1975: Hokô-robotto to Jinkô no Ashi (Legged Locomotion Robots). Translated by Katô, I. & T. Yamashita, Tokyo. (Japanese)
Wiener, N. [1950] 1954: The Human Use of Human Beings. Cybernetics and Society, Garden City (NY).
Yamamoto, K., M. Arai, R. Ichinose, & Y. Ono 2011: "Hitachi no robotto kaihatsu (Development of robots at Hitachi)." In: Proceedings of the 29th Annual Conference of the Robotics Society of Japan, 2F1-1. (Japanese)
Yin, Y., Y. Nakashima & T. Yoshikawa 2010: "Simulation ga sasaeru pâtonâ robotto kaihatsu (Development and application of simulation technology to Toyota partner robots)." In: Proceedings of the 28th Annual Conference of the Robotics Society of Japan, 2E1-6. (Japanese)

Acknowledgements

This study is supported by the Japan Society for the Promotion of Science (JSPS), KAKENHI (Grants-in-Aid for Scientific Research) No. 21520004, and the Global COE (Center of Excellence) Program *The University of Tokyo Center for Philosophy (UTCP)* of the Ministry of Education, Culture, Sports, Science, and Technology, Japan. Professor Atsuo Takanishi at Waseda University gave me some documents and information on the history of robotics at Waseda University.

Robotic Appearances and Forms of Life. A Phenomenological-Hermeneutical Approach to the Relation between Robotics and Culture

Mark Coeckelbergh

Introduction

Cultural differences with regard to how people relate to robots raise questions concerning the nature, causes, and meaning of these differences. How can we conceptualize the relation between robotics and culture? Focusing on differences between the West and Japan in the perception and design of human-like intelligent autonomous robots, this paper discusses different approaches to the relation between robotics and culture. It is argued that the limitations of realist, dualist, and objectivist scientific methodologies can be overcome if we adopt a hermeneutical-phenomenological approach, which understands the usual "cultural" and "historical" explanations of differences in robotic culture not as scientific explanations, but as part of a hermeneutical process and as illustrative of different forms of life. Robots are revealed as hermeneutic tools that function within techno-anthropologies, for example Western negative anthropologies.

By reflecting on robot ontology, meaning and appearance, the paper suggests a non-dualist view that reveals robots as hermeneutic tools that contribute to our self-understanding as humans. It also employs a transcendentalist argument in order to show that different ways of viewing and developing robots are only possible on the basis of material-social forms of life, which may differ between East and West. However, forms of life do not only function as conditions of possibility for the perception and design of robots; to some degree they are themselves also shaped by robotic design, use, interaction, experience, and discourse. This analysis attends us to the hermeneutic-phenomenological responsibility of robotics designers, engineers, managers, and policy makers in East and West. It recommends an ethics that demands more than adherence to safety guidelines, deontological codes, and ethical principles: it asks us to imagine how particular robots may fit into a particular form of life and how they might contribute to ongoing transformations of that form of life.

It is often suggested in the media that the Japanese love robots, whereas the West fears them. For example, an article in "The Times" headlines: "Japan: The Nation that Loves Robots" and claims that "to say that the Japanese are fervent about robots is a great understatement" (Lewis 2009). And Hornyak´s book on Japanese robotics carries the title "Loving the Machine" (Hornyak 2006). But while there is little doubt that robots are popular in Japan, it remains unclear if there really is a difference in attitude between Japan and the West, what the extent and nature of this difference is, how this difference can be explained, and what it means for robotics use and design. More generally, the issue of the "Japanese difference" raises the deeper question concerning the relation between technology and culture: Is the use and design of technology dependent on culture, and if so, in what way?

Philosophy of robotics can contribute to discussions about this issue by helping to conceptualize the relation between robotics and culture. Focusing on differences between the West and Japan in the perception and design of human-like intelligent autonomous robots, this paper will discuss different approaches to the relation between robotics and culture. It will be argued that the limitations of realist, dualist, and objectivist scientific methodologies can be overcome if we adopt a hermeneutical-phenomenological approach.

First I will present some common explanations of cultural differences between Japan and the West, featuring the history of robotics in Japan, Japanese popular culture, Japanese religion and world-views, and Japanese society and ethics. Then I will criticize these "cultural" and "historical" explanations, at least in so far as they pretend to be scientific explanations of differences in attitude. I will argue that they should rather be understood as part of a hermeneutical process in which we try to understand ourselves as humans and as illustrative of different forms of life, which function as conditions of possibility for robotic use, design, and discourse.

From Robots to Histories, Cultures and Gardens

The usual answer to the question why the Japanese are more inclined to adopt and accept humanoid and social robots tells a story about puppets, mangas, gardens, and spirits (see for example Schodt 1988; Wagner 2009).

References are often made to the history of robotics in Japan. In the Edo period (17th-19th-century) *Karakuri* puppets were used: automata with mechanical parts inside, inspired by Chinese and Portugese designs. A *Karakuri* automaton would be able to carry tea and make other performances. Appearance and show mattered: "*Karakuri Ningyo*" (mechanized puppet) means a mechanical device to trick or surprise (*Karakuri*) that has a person-shape (*Ningyo*) (Law 1997; Boyle 2008). Moreover, Japan is also known for its history of *Bunraku*, a tradition of puppet theatre which started in Osaka in the 17th-century and has been continued ever since. Hornyak also mentions Nishimura´s "Buddha robot": a bronze robot created for the 1928 Kyoto Fair (Hornyak 2006).

This history of Japanese popular culture is also often mentioned to account for the popularity of robots in Japan. It has been argued that in the past popular mangas like Tetsuan Atom (Astro Boy) have helped to construct a positive view of robots as friends or companions that live side by side in harmony with humans (or are in any case not evil). Contemporary robotics also seems to put the emphasis on entertainment and companionship (for example with Aibo, Asimo,

Paro, and other friendly, often animal-like robots). This is then contrasted with Western negative views of robots. Wagner et al. write:

> "We find a tension between the Japanese idea of advanced technology as beneficent helper, companion, even savior, and the Western apprehensiveness that advanced technology could become an alien, threatening, and destructive force."

The Japanese view of nature and its relation to technology has also been mentioned as an explanation for the high acceptance of social robots in Japan: as Japanese gardens show, the aim is not so much to control nature but to imitate it. This imitation is artificial, of course, but this is not seen as problematic. The "artificial reproduction of nature" (Kaplan 2004) is a homage to nature, up to the point of imitating mistakes (Sone 2008). Kitsch is not problematic since the artificial/natural distinction lacks significance in this sense. This seems to render robots less problematic.

Furthermore, explanations of robot popularity in Japan usually make references to Japanese religions and world-view(s). Whereas the Western view presupposes the existence of a self and is dualist and individualist, the three religious (under)currents in Japanese culture entertain different views. Buddhism rejects the idea of a (real, stable) self. Shintoism holds that spirits can inhabit objects, animals, trees, and rocks. Even tools have spirits, which live in harmony with human beings and are taken to the temple when broken (Kitano 2006). Robots are therefore taken to be "living things" and we can live with them since "nature is not external to culture and society" (Robertson 2010). Japan is also influenced by Chinese philosophy, in particular Confucianism and Daoïsm, which teach that the self is "interdependent" and must be understood in a relational way (Lai 2007). Relations with robots, it seems, are therefore unproblematic.

Finally, Japanese society is collectivist and emphasizes social harmony, consensus, trust, loyalty, and group responsibility (Herbig & Palumbo 1994; Kitano 2006; Wagner et al. 2005; Saha 1994). Although this emphasis on the group also has downsides – it is often called "ethnocentric" (Herbig & Palumbo 1994, p. 93) – it means that its ethics is a social ethics as opposed to the Western emphasis on individual responsibilty. Ethics (Rinri) is equated with "the study of the community, or of the way of achieving harmony in human relationships" (Kitano 2006, p. 80) and is "defined by the relationship between the individual and the community to which he/she belongs" (Nishigaki 2006, p. 238). Tetsuro Watsuji, a 20th-century Japanese philosopher, has written a book on ethics called "*Rinri-gaku*" (for a partial translation in English see Watsuji 1996), which literally means the study of the "betweenness" of human beings (Kitano 2006, p. 82). What matters is living together in a harmonious community. With or without robots.

From Facts and Explanations to Interpretations and Constructions of Meaning

However interesting, it remains doubtful if these references to Japanese history and culture can count as an explanation of the popularity of robots in Japan as opposed to the West – if there is a significant difference in attitude at all.

First, is the attitude towards robots really different? A study by Bartneck et al. (2007) showed that the Japanese are not as positive as stereotypically assumed. And MacDorman et al. (2009) found similarities in attitude among faculty of US and Japanese universities.

Second, Western popular culture has friendly robot characters too, both in fiction (R2D2, C-3PO, Wall-E) and in reality (for example the social robots developed by Breazeal´s team at MIT, US).

Third, not only culture but economic and political factors play a role as well, such as the expertise with factory automatisation and the search for additional markets (MacDorman et al. 2009), a cooperative model of innovation (Saha 1994), and post-war governmental pressure to develop robots for entertainment and for solving the demographic crisis rather than for military applications.

Fourth, can we really establish a causal relationship between on the one hand culture and on the other hand the development and adoption and technology, as Herbig and Palumbo (1994) suggest? Not only are empirical data still missing, as Wagner (2009) says, but the very project of trying to provide proof for causal relationships in this domain is at least very questionable.

Finally, these constructions of robotics history are vulnerable to the charge of orientalism and self-orientalism. Do the Japanese really have a "natural" relation to robots? For example, some argue that there are also animistic tendencies in the West (Aupers 2002). And as Wagner has argued, the construction of a continuous line of robot loving has been used by the Japanese to justify the technological empowerment of post-war Japan and a positive image of robots helped to give hope to the people and made it easier to accept robots rather than foreign workers as a solution for labor shortage (Wagner 2009, pp. 511-514). Thus, it seems that robotic (self-)interpretations are important to both East and West. This is especially so given the relation between technological development and globalization. As Nishigaki says:

> "Japan is by no means the only country in the world that is faced with the collapse of traditional communities and the unsettling of ethical and moral values caused by the rapid penetration of information technologies into every corner of society and by the globalization of economic activities." (Nishigaki 2006, p. 242)

On these grounds, I propose to regard references to the history and culture of Japan not as scientific explanations of differences in attitudes towards robotics, but as part of ongoing hermeneutic exercises – in East and West – that aim at interpretation, understanding and self-understanding rather than explanation, and are concerned with meaning rather than facts. Viewed in this light, the previous overview of historical and cultural elements should not be understood as a collection of facts and factors, but as hermeneutic exercises: constructions and interpretations of history and culture that help us to come to terms with techno-social developments in robotics. At the same time, robots are used to construct that culture. As I will argue below they are not mere material objects that stand completely apart from culture; they also structure our experiences and beliefs.

From this hermeneutic perspective, it is interesting to look not only at Eastern robotic culture, but also at the Western attitude towards robots and the culture it presupposes and shapes.

It has been claimed that robots challenge the Western world-view and sense of identity. Robots present what has been called a "category boundary problem" (MacDorman & Cowly 2006; Ramey 2005; Turkle 2007) since they lie on the boundary between human and non-human. In contrast to Japan, so it is claimed, this does pose a challenge to the Western sense of personal and human identity (MacDorman et al 2009, p. 487), since in the West people think it is important to make that human/non-human distinction. We want to be special. Kaplan has even argued that this is true for technology in general: in the West technology seems "fundamental for defining what humans are" (Kaplan 2004).

But is this only true for the West? It seems that robots, and perhaps other technologies also, function in West and East as what I propose to call "hermeneutic tools": we use them for defining and re-defining the human. They function as mirrors that help us to understand ourselves and the world around us. They function within a cultural-hermeneutical process. More generally, the way we define ourselves as humans depends on the observation, construction, and imagination of quasi-humans or non-humans.

In the West, we have "used" gods, angels, demons, animals (e.g. apes), "Golems," "Homunculi," machines, monsters, computers, artificially intelligent systems, "Zombies," "Aliens," and robots as hermeneutical tools to construct the human. For example, in the Jewish tradition the "Golem" is said to be created by man as an imitation of divine creation, although this does not mean that it is the same as divine creation. In this way, "Golem-stories" serve to explore both similarities and differences between human and divine creation, and between humans and God. And Descartes´ philosophical anthropology, including the famous mind-body distinction (or rather: soul-body distinction), heavily relies on the construction of animals as complex machines and indeed on the construction of the human body as a machine. (Self-definitions usually involve selective and one-sided constructions of quasi-humans or non-humans.) Without using animals and machines, Descartes could not have developed his view; he needed them in his (early-) modern hermeneutics.

Typically, Western self-definitions come in the form of what I propose to call a "negative anthropology": in the West humans are defined by what they are not. They are not-gods, not-animals, not-"Zombies," and indeed not-robots. In this way, robots and (some) other technologies provide a "via negative" that helps us to define ourselves as humans. New and emerging technologies like robots are used to explore the boundaries of the human. In the past it was held important to distinguish ourselves from the gods and from God. In the history of Western culture we find the Romantic idea that we are alienated from our nature and the idea that we should not play god (or God) since that would constitute hybris. This has involved a Romantic re-interpretation of ancient stories (the tower of Babel, Goethe´s "Der Zauberlehrling," "Golem-stories," "myth of *Prometheus*," "Faust," etc.). In contemporary times it is more important to distinguish ourselves from technology, since it is held that technology has caused alienation and encouraged hybris. This gave rise to the so-called "Frankenstein syndrome": the idea that if we play God by means of technology, this may turn against us; technology will dominate us rather than the other way around (see for example Čapek´s play "R.U.R.," which is about a revolt of robots, Asimov´s Three Laws of Robotics as a response to this danger, "Terminator," the film "I, Robot" etc.). Contemporary Western culture, in particular science-fiction but also contemporary art, explores the differences between robots and humans: Are we more than machines?

Thus, in both Western and Eastern culture we can observe that the meaning and boundaries of the human are explored and constructed by using machines in general, and robots in particular. Robots are hermeneutic tools within what we may call "techno-anthropologies:" they help us to define the (boundaries of the) human.

From Reality to Appearance and Forms of Life

This approach to robots can be further clarified and developed by using a phenomenological perspective.

Standard robot ontology is realist and objectivist: a distinction is made between on the one hand "what the robot really is" (a machine, implementation of code, information, etc.: "objective reality") and on the other hand the appearance of the robot ("subjective," depends on human perception). Hence the engineer´s task is to design for appearance (to produce a certain appearance, to be a "master of illusion") and the scientist´s task is to unmask, reveal, uncover, strip away the phenomena in order to show objective reality, purified from human subjectivity, perception, and emotions. (An example of this approach can be found in Kahn et al. 2006, who categorize claims about human-like robots in a way that distinguishes between "people´s psychological beliefs and actions" and "the correct ontological status of the robot" (Kahn et al. 2006, p. 365).)

But can we know the "correct" ontological status of anything? And is it always meaningful to discuss ontological differences between humans and robots? As Kitano writes, in Japan "there is unlikely to be much philosophical discussion about, for instance, 'what robots are and what humans are' as takes place in the West" (Kitano 2006, p. 79). In the West, however, such discussions are central. This may have to do with Western metaphysics, which is dualist: since Plato a distinction is made between the world of appearance and the real world, between the subjective and the objective. Is there a way to go beyond standard ontology and, more generally, beyond Western metaphysics?

An obvious way to proceed here would be to turn to non-dualist Eastern world-views in order to explore an alternative robot ontology. However, there is a route available to us that is much closer to home: the tradition of (Western) phenomenology, which has always challenged realist, objectvist, and dualist metaphysics. Phenomenological and hermeneutical approaches to technology (ranging from Dreyfus and Ihde in the U.S. to Irrgang in Germany) draw on Heideggerian phenomenology to show that there is more than one way of seeing artifacts and that they can have more than one meaning. For example, Ihde has argued that the meaning of technological artifacts are "multistable" (Ihde 1990). Hence a robot can appear to us as a human or a "quasi-other" (Ihde 1990) and as a machine. This means that there is not one, "objective" reality but rather something like "Gestalt": the same object (to call it an object is only one way of seeing it) can be viewed in entirely different ways. Our perception of a particular robot is mediated by our ways of seeing. This means that there is not one "correct" way of seeing. Any "correctness" depends on what we, as embodied and inner-wordly beings, perceive from within a perceptual scheme that is available in a particular culture. However, the robot is not "interpreted," since this way of speaking remains dualist in so far it assumes that "first" there is objective reality and "then" (cultural) interpretation. Rather, the robot already appears to us in a particular way, is already interpreted. This process does not just depend on us as individuals (on our individual will), but also on the culture and social context we live in and on the non-linguistic, embodied and skilled ways of perceiving and doing that are nurtured by that culture and co-constitute that culture, on the structures of what Heidegger called "being-in-the-world."

I propose to articulate this phenomenological-hermeneutical alternative to objectivist thinking in the form of a transcendentalist argument: how robots appear to us depends on cultural and bodily conditions of possibility that both enable and limit how we can view these robots. Thus, the appearance of the robot depends on preconditions that could be called "forms of life" (Wittgenstein). This means that the mentioned historical and cultural differences do not really cause attitudes towards robots, but are rather to be regarded as descriptions of forms of life: forms that enable and limit what we say about robots and how we act towards robots. They are contributions to a robot hermeneutics, which itself depends on the forms of life it tries to describe.

This approach has not only implications for what can be said about robots, but also what can be said about humans. In the light of the hermeneutical approach, anthropology becomes a hermeneutics of the human that is necessarily a relational and non-dualist anthropology: humans (formerly called "subject") and robot (formerly called "object") mutually constitute each other. In this way, a dualist ontology and metaphysics is overcome. Put in transcendentalist language, this approach suggests that our human-talk, like our robot-talk, depends on a relational a priori – relations between humans, robots, things, etc. – which structures our ways of speaking, perceiving, and doing. In other words, with regard to anthropology both the hermeneutics and its preconditions (the forms of life) must be understood in relational terms.

This view seems compatible with relational and non-dualist Eastern philosophy. For example, the influential Japanese philosopher Nishida Kitaro (1870-1945) has a non-dualist notion of pure experience that is prior to all reflection (Nishida 1990; Feenberg & Arisaka 1990), which is the ontological basis of reality (Feenberg 1994), perhaps even "the always already present ground even of reflection itself" (Feenberg 1999). The forms of life in question seem to constitute such a ground. There may also be parallels to non-dualist metaphysics in Buddhism, Taoism, and Confucianism. However, I will not further develop this comparative inquiry here.

Summary and Conclusion for Responsibility

In this paper I have shown that a phenomenological-hermeneutical approach can conceptualize the precise relation between robotics and culture in a novel way. I have argued that common descriptions and explanations of cultural differences between the "West" and Japan with regard to robots should not be understood as causal explanations of differences in attitude, but rather as attempts to understand both robots and the human. By reflecting on robot ontology, meaning and appearance, I have suggested a non-dualist view that reveals robots as hermeneutic tools that contribute to our self-understanding as humans. I have employed a transcendentalist argument in order to show that different ways of viewing and developing robots are only possible on the basis of material-social forms of life, which may differ between East and West. This material, technological aspect is important: forms of life do not only function as conditions of possibility for the perception and design of robots; to some degree they are themselves also shaped by robotic design, use, interaction, experience, and discourse.

This analysis attends us to the hermeneutic-phenomenological responsibility of robotics designers, engineers, managers, and policy makers in East and West. It suggests an ethics that demands more than adherence to safety guidelines, deontological codes, and ethical principles: it asks us to imagine how particular robots may fit into particular forms of life and how they might contribute to ongoing transformations of those cultural-technological life-forms, which ground and structure what we call real, valuable, and meaningful.

References

Aupers, St. 2002: "The Revenge of the Machines: On Modernity, Digital Technology and Animism." In: Asian Journal of Social Science 30 (2), pp. 199-220.

Bartneck, Ch., T. Suzuki, T. Kanda & T. Nomura 2007: "The Influence of People´s Culture and Prior Experiences with Aibo on their Attitude towards Robots." In: AI & Society 21 (1-2), pp. 217-230.

Boyle, K. 2008: Karakuri-Info. online: http://www.karakuri.info/

Feenberg, A. 1994: "The Problem of Modernity in the Philosophy of Nishida." In: Heisig, J. & J. Maraldo (eds.) 1994: Rude Awakenings, Honolulu.

Feenberg, A. 1999: "Experience and Culture: Nishida´s Path 'To the Things Themselves.'" In: Philosophy East & West 49 (1).

Feenberg, A. & Y. Arisaka 1990: "Experiential Ontology: The Origins of the Nishida Philosophy in the Doctrine of Pure Experience." In: International Philosophical Quarterly, 30 (2), pp. 173-204.

Herbig, P. A. & Palumbo, F. 1994: "The Effect of Culture on the Adoption Process: A Comparison of Japanese and American Behavior." In: Technological Forecasting and Social Change 46 (1), pp. 71-101.

Hornyak, T. N. 2006: Loving the Machine: The Art and Science of Japanese Robots, Tokyo & New York.

Ihde, D. 1990: Technology and the Lifeworld: From Garden to Earth, Bloomington.

Kahn, P. H., H. Ishiguro, B. Friedman & T. Kanda 2006: "What is a Human? Toward Psychological Benchmarks in the Field of Human-Robot Interaction." In: Proceedings of the 15th IEEE international symposium on robot and human interactive communication. Hatfield, UK, September 6-9, pp. 364-371.

Kaplan, F. 2004: "Who is Afraid of the Humanoid? Investigating Cultural Differences in the Acceptance of Robots." In: International journal of humanoid robotics 1(3), pp. 465-480.

Kitano, N. 2006: "'Rinri:' An Incitement towards the Existence of Robots in Japanese Society." In: International Review of Information Ethics 6 (12), pp. 78-83.

Lai, K. L. 2007: "Understanding Change: The Interdependent Self in its Environment." In: Journal of Chinese Philosophy 34 (suppl. s1), pp. 81-99.

Law, J. M. 1997: Puppets of Nostalgia, Princeton.

Lewis, L. 2009: "Japan: The Nation That Loves Robots." In: The Times, February 27, 2009.

MacDorman, K. F. & St. J. Cowley 2006: "Long-term relationships as a benchmark for robot personhood." In: Proceedings of the 15th IEEE international symposium on robot and human interactive communication. Hatfield, UK, September 6-9, pp. 378-383.

MacDorman, K. F., S. K. Vasudevan, & Chin-Chang Ho 2009: "Does Japan Really Have Robot Mania?" In: AI & Society 23 (4), pp. 485-510.

Nishida, K. 1990: An Inquiry into The Good. (trans. M. Abe & Ch. Ives), New Haven & London.

Nishigaki, T. 2006: "The Ethics in Japanese Information Society: Consideration on Francisco Varela´s The Embodied Mind from the Perspective of Fundamental Informatics." In: Ethics & Information Technology 8(4), pp. 237-242.

Ramey, Ch. H. 2005: "The Uncanny Valley of Similarities Concerning Abortion, Baldness, Heaps of Sand, and Humanlike Robots." In: Proceedings of the "Views of the Uncanny Valley" Workshop, IEEE-RAS International Conference on Humanoid Robots,

Tsukuba, Japan.

Robertson, J. 2010: "Gendering Humanoid Robots: Robo-Sexism in Japan." In: Body & Society 16 (2), pp. 1-36.

Saha, A. 1994: "Culture and the Development of Technology in Japan." In: Technology in Society 16 (2), pp. 225-241.

Schodt, F. L. 1988. Inside the Robot Kingdom: Japan, Mechatronics, and the Coming Robotopia, Tokyo & New York.

Sone, Y. 2008: "Realism of the Unreal: The Japanese Robot and the Performance of Representation." In: Visual Communication 7(3), pp. 345-362.

Turkle, Sh. 2007: "Authenticity in the Age of Digital Companions." In: Interaction Studies 8 (3), pp. 501-517.

Wagner, C. 2009: "'The Japanese Way of Robotics': Interacting 'Naturally' with Robots as a National Character?" In: Proceedings of the 18th IEEE International Symposium on Robots and Human Interactive Communications, 27.9.-2.10.2009, Toyama/Japan, pp. 169-174.

Wagner, J. J., D. M. Cannon & H. F. M. Van der Loos 2005: "Cross-Cultural Considerations in Establishing Roboethics for Neuro-Robot Applications." In: Proceedings of the 2005 IEEE 9th International Conference on Rehabilitation Robotics, June 28 - July 1, 2005, Chicago, IL, USA.

Watsuji, T. 1996: Watsuji Tetsurō's Rinrigaku: Ethics in Japan trans. from the First Half of Rinrigaku Vol. 1. (trans. S. Yamamoto & R. E. Carter), Albany.

Humanoid Robots and Human Knowing – Perspectivity and Hermeneutics in Terms of Material Culture

Michael Funk

Introduction

In this approach I am going to combine the philosophy of understanding and interpreting (hermeneutics) with the philosophy of knowing (epistemology). Related to a concept of perspectivity emphasis is on robots in terms of material culture. Human knowing is introduced as a process of tacit understanding and of interrelating perspectivity with forms of knowledge in particular situations. Those situations are shaped by cultural horizons: 1PPP ("first-person-perspective-plural": native culture), 2PPP ("second-person-perspective-plural": foreign culture)[1] and 3PPP ("third-person-perspective-plural": transcultural objective perspective related to natural laws or ethical argumentations, which is also applied in the form of political or legal organizations and Technology Assessment[2]). Moreover the perspectivity is also related to: 1PP ("I"), 2PP ("You") and 3PP (external observer). While we are interrelating the personal actions (1PP) with social feedback (2PP), we are always involved in a concrete cultural situation (language, gestures, norms, values... [1PPP]). Within those processes, that is my hypothesis, we intertwine five forms of knowledge: 1. sensorimotor movements; 2. perceptions; 3. emotions; 4. propositions; and 5. philosophical-reflections[3]. What does this mean for "Humanoid Robots and Human Knowing"?

The basic assumption is that "humanoid robots" are robots with a human-like appearance (Coeckelbergh 2011a, p. 62; Coeckelbergh 2011b, p. 199; Christaller et al. 2001, pp. 87-88; Decker 2010, p. 45; Hirukawa 2007, p. 1; Knoll & Christaller 2003, pp. 12-14). However, does that entail that "humanoid robots" have human knowledge as well? I do not think so. If we believed that humanoid robots have human knowledge, we would belief that human knowledge could be reduced to only one form, namely propositional knowledge

1 Bernhard Irrgang elaborates a concept of perspectivity that is reframing the 2PPP as "institutions" (Irrgang 2009a, p. 109). From my point of view, institutions are related to the 3PPP and the 2PPP belongs to the framework of different cultures.

2 See also the chapter "Who is taking over? Technology Assessment of Autonomous (Service) Robots" (Michael Decker) in this book. In Technology Assessment, another concept related to disciplinary perspectives is elaborated. Because of the multidisciplinary challenge, the term "perspective" is used to illustrate the different disciplinary approaches to robotics: "Technological perspective," "Economic perspective," "Legal perspective," "Psychological perspective," "Philosophical and ethical perspectives" (Decker et al. 2011, pp. 38ff.; see also Decker 2011, pp. 250-253; Christaller et al. 2001, pp. 210ff.).

3 Philosophical reflexive knowledge is typical for human beings: a matter of practical success and of logical truth (Irrgang 2005b, p. 247).

related to the 3PP within the 3PPP (natural and technical laws that are seen as interculturally valid). Robotics research is strongly motivated by an attempt to figure out who we are, while trying to create a copy of us (Decker 2010, pp. 48-49; Irrgang 2005a, p. 167; Kemp et al. 2008, p. 1308).[4] And those attempts are usually driven by mathematical and physical approaches (Christaller 2007, p. ix; Janich 1999, p. 18; Kajita 2007, p. xi), which shape the modern sciences and engineering sciences since Bacon, Descartes, Galileo, Kepler or Newton. In terms of robotics: human body-movements and perceptions in copper, silicon and iron; human cognition as 0 and 1. And at the same time, our attempts to build (humanoid) robots are characterized by implicit anthropological assumptions (Decker 2010, p. 42). Insofar, Bionics could be a fruitful paradigm in robotics-research (Christaller et al. 2001, p. 25, p. 74; Decker 2010, p. 48), as long as we do not attempt ourselves to behave like copper and iron in 0 and 1.

The genetic and *leiblich*-sensory processes of human organic knowing are much more complex and there is a knowledge gap as well: we are not able to transform every piece of tacit knowing in the propositional form of the 3PP and 3PPP; this is the reason why robots cannot replace human knowing. And they cannot replace human creativity because of the same reason: human knowing is a process of interrelating the perspectives with the five forms of knowledge in contingent situations. If we could grasp unforeseeable situations from the 3PP or the 3PPP, quasi as external observers, robots could do the same, but:

> "Would it be possible to design an experimental robot that could substitute this creative human action 'in an equivalent, like-for like way?' Evidently it would not, because in order to be able to design a machine of this kind it would be necessary to foresee all the unforeseeable incidents [...] attempting to foresee rational goal-driven responses to unforeseeable incidents is attempting the impossible." (Janich 2012, p. 219)

And:

> "'Intelligence' embedded in machines; embedded in biological organisms and embedded in human bodies (Leiber) is always something different. Animals have merely a 1st-person-perspective, robots have no perspectivity." (Irrgang 2008a, p. 62)

On the other hand – and here we come to material culture in terms of technical tools and their social implementation since the first hand axes – the praxis of interacting with robots will influence human processes of knowing. This could be called a "human-robot-mirror." Mirror means that robots will not replace but enhance our 2PP as we are interacting with them, which also includes a verbal level.[5] Robots talk to us and we are talking to robots.[6] They could shape our human

---

4 This is one of the following three reasons for constructing "humanoid robots": 1. the old dream to build artificial humans; 2. to learn something about ourselves; and 3. because of purposive, means-end rationality (Decker 2010, p. 46; see also Coeckelbergh 2011b, p. 200).

5 That has been discussed by Don Ihde and Mark Coeckelbergh as "quasi-other" (Ihde

sensorimotor and perceptual knowledge and the relation between 1PP and 2PP. And even our emotional knowledge may be involved in those processes. Robots will not replace human beings as such but they can be seen as another technological piece of material culture that enhances and challenges the perspective praxis and interactions in human (social) life. In the next sections I am going to elaborate the methodological and philosophical framework of this approach.

Hermeneutics, Pragmatic Phenomenology and Implicit Knowing

Two basic concepts of philosophical hermeneutics can be differentiated. The first one is a classical notion that is related to interpreting texts or other linguistic structures. Here we can draw a line beginning with forms of theological hermeneutics, which is the art and science of interpreting the bible, up to 19th-century methodological movements by Friedrich Schleiermacher. He is reframing hermeneutics as a text-based more general methodology for the *Geisteswissenschaften*, the social sciences and philosophy (Joisten 2009, pp. 17-18, pp. 95ff.). This first classical notion can be called "linguistic hermeneutics," which is continued in 20th-century approaches, e.g. Hans-Georg Gadamer's (Ibid., pp. 149ff.).

Somehow the 19th-century plays an important role in the development of scientific methodology insofar, as it fueled a defensive behavior in the attitude of philosophical thought. Maybe Leibniz can be seen as one of the last universally educated scholars. In 19th-century developments a methodological gap between the natural sciences and philosophy started to take shape: positivism and scientism on the one side; (linguistic) hermeneutics on the other. After the disappearance of the last polymaths (and today's world is much more complex, we should not wait for any comeback) and the rise of positivistic thinking, philoso-

---

1990, pp. 97-108, p. 107; Coeckelbergh 2011a, p. 61, p. 65; Coeckelbergh 2011b, p. 198). The mirror-effect is not only reserved to human-robot-interactions but has its roots in human-human-interactions: "Or, perhaps seeing other humans and recognizing their shapes, we refer back to ourselves with the other-as-mirror." (Ihde 2010, p. 42) "The robot becomes a 'you'-not as a stand-in for someone else (which we might call a 'delegated second person') but a 'you' in its own right, an artificial second-person, which has a claim on me as a social being. […]; by talking to the robot in second-person terms, they [people] also construct it as a quasi-other." (Coeckelbergh 2011a, p. 65)

6 The important epistemological point is incorporated into the thesis that human communicative competence is principally not substitutable by technical artifacts and systems (Janich 1999, p. 19). But "it becomes evident that human-like robots may have a profound impact on the nature of our communication" (Nishida 2009, p. 107; see also Gill 2008). "The core capability for some service and personal robots is to interact with the user in a natural language." (Nishida 2009, p. 108) "In current human-robot relations, we can observe a shift from talking about robots and about human-robot relations to talking to robots. Let me [Mark Coeckelbergh] explain this shift and bring out its linguistic dimension and philosophical significance by distinguishing between different 'perspectives.'" (Coeckelbergh 2011a, p. 63)

phers started searching for a strategy to legitimize their position in a world of independent natural and technical sciences. The narrow notion of hermeneutics as "linguistic hermeneutics" is one fruit of that movement, which is caused by an epistemic and social emancipation of the natural sciences and the technically oriented engineering sciences.

As soon as the natural and technical sciences drifted away from philosophy, philosophers started understanding hermeneutics as a non-technical and non-natural-scientific methodology. The narrowing of hermeneutics as linguistically oriented text-hermeneutics after Schleiermacher or Gadamer can be illustrated as one aspect of this philosophical defensive movement. As soon as this narrow hermeneutics becomes a "*hermeneutische Weltanschauung*" (hermeneutical worldview), it loses sight for the natural sciences and their genuine justifications and praxis (Jung 2002, p. 134). Leibniz was a philosopher and a technician in personal union. But if we talk about philosophy of robotics or roboethics today, we find a gap: engineers and their knowing in developing, constructing or repairing robots on the one side; and philosophers reflecting the aspects of technical praxis or social and cultural implementation with respect to ethics on the other.[7] As long as hermeneutics remains text-oriented and a "linguistic hermeneutics," philosophy is not well-prepared to provide a contribution for bridging this gap.

In a general sense, we can say that understanding is related to active processes of repetition and changing positions, i.e. changing the perspective. This starts with the first body movements of babies. Insofar hermeneutics is more than interpreting texts; it is a knowledge-based process of interfacing multiple perspectives (Jung 2002, pp. 133-134, p.148; Joisten 2009, pp. 199-202), a form of understanding the world around us and last but not least understanding who we are.

Specific human forms of knowing are related to what could be labeled "pragmatic hermeneutics" (Jung 2002, p. 135, pp. 143ff., pp. 151ff., p. 159; Joisten 2009, p. 200; Kurthen 1994, p. 13). And it is at this point where we can find methodological insights that lead to a concept of philosophical hermeneutics, which is applicable to questions of human-robot-interactions: the amalgamation of hermeneutics and pragmatic phenomenology in the early 20th-century. As far as I can see, the first steps of phenomenology after Clemens Brentano and the early Edmund Husserl where still characterized by defensiveness. But on the other hand, the later developments of phenomenology led to some fruitful so-called "phenomenological-hermeneutical" approaches, which bear the potential to overcome a narrow and defensive form of hermeneutics. In the 1930s, Edmund Husserl developed a new form of phenomenology in his "Crisis"-book (Husserl 2012), which is close to what today is called "postphenomenology"

7 A third form of expertise belongs to the way we know how to use robots in everyday life. For this, it should not be necessary to have studied philosophy or an engineering science.

(Ihde 1993) or "transclassical phenomenology" (Irrgang 2009a, pp. 83-182). Don Ihde argues for an epistemic shift, which arises within the Philosophy of John Dewey, Edmund Husserl, Maurice Merleau-Ponty or Hubert Dreyfus. For René Descartes, the main epistemic reason had been consciousness. But

> "[...] I [Don Ihde] am trying to show how embodiment replaces subjectivity – at least the 'mind' subjectivity of early modern epistemology – in a now modified postphenomenology." (Ihde 2010, p. 42)

In his concept of "*leiblicher Geist*" (embodied mind) Bernhard Irrgang argues for the same epistemic shift (Irrgang 2007, pp. 184ff.; Irrgang 2009a, pp. 61ff.). His main hypothesis is that mind is not a matter of consciousness, but of competence (Irrgang 2009a, p. 61). Irrgang′s central claim in epistemology is the link between embodiment, technical understanding and a specific kind of embodied knowledge. Therefore he introduces the notion of "*Umgangswissen*;" handling-knowledge, which is strongly associated with Michel Polanyi′s concept of "implicit" and "tacit knowing"[8] (Irrgang 2005a, pp. 43-46, p. 138).

Husserl did not only develop the roots of a post-Cartesian phenomenology that replaced the mathematical transcendental and disembodied subject by a subject of bodily actions in a concrete lifeworld. He also established an approach of non-linguistic hermeneutics. The second author in that philosophical trajectory is Martin Heidegger with his early works including "Being and Time" and the third one is Maurice Merlau-Ponty. These authors are important roots for what is now called "expanded hermeneutics" (Ihde 1998), "material hermeneutics" (Verbeek 2005, pp. 121-146) or "technical hermeneutics" ("*Technikhermeneutik*") (Irrgang 2009b). Hermeneutics is more than interpreting texts, moreover it involves processes of understanding who we are, and bodily (*leibliche*) processes of sensory and perceptual understanding and interpretation. Materiality in both meanings – technologies and the culturally shaped human body – is involved. Here is the point, where hermeneutics and epistemology are intertwined. The same shift that happened within 20th-century philosophical hermeneutics happened to philosophical epistemology as well; a turn away from the narrow focus on linguistic-semantic structures, semiotics or language in the sense of formal logic and symbols. Not only propositional knowledge, but primarily the tacit fundamentals of human knowing gained the epistemological center stage. The empirical and embodied turn of hermeneutics is related to a pragmatic and embodied turn in epistemology[9] and an "empirical turn" in the philosophy of

---

8 "I shall reconsider human knowledge by starting from the fact that we can know more than we can tell." (Polanyi 2009, p. 4; see also Polanyi 1969; Polanyi 2009)

9 "Meansend knowledge, and knowledge for correcting faults can only be referred to action, not to behavior." (Grunwald 2012, p. 193) And that is the reason why a hermeneutics of technologies is a pragmatic one. Success or failure are pragmatic truth-criteria for embodied knowing (tacit- or implicit knowledge and *Umgangswissen*). The ground for knowledge in human actions is not logical evidence but pragmatic success (Irgang 2005b,

technology (Achterhuis 2001) as well. According to these approaches, which are close to Michael Polanyi´s concept of implicit knowledge (Mahrenholz 2011, pp. 242-243) and supported by an embodied turn in the cognitive sciences (Johnson 1987; Varela et al. 1997, pp. 172f.; Noë 2004), technical praxis, gestures, and the use of handcraft tools, sensory body movements and perceptions have been accepted as genuine non-linguistic but pragmatic forms of human knowing (Dreyfus 1979; Heidegger 2006, §§ 14ff.; Ihde 1998; Irrgang 2009a, p. 188ff.; Piaget 1954; Rentsch 2003, p. 15; Ryle 1946; Wittgenstein 2006, § 78[10]). There is a genetic-organic basis for that on the one side; and a cultural on the other.[11] Human knowing is a process of realizing genetic-organic potentials within concrete cultural horizons while interacting with the social-human and material-cultural environment.

This can be illustrated with the example of music. Linguistic hermeneutics is related to interpreting texts, and that includes musical scores. If you see a piece of composed music, you may understand it terms of propositions or notational harmonic structures and say something like: "I know that Beethoven has composed the first movement of Op. 27.2 – the moonlight-sonata – in c-sharp minor." What is being interpreted in this context is the semantic meaning of four specific symbols: the four crosses at the beginning of the musical notation. They belong to E major as well as to C-sharp minor. But the context and the first octave of the left hand are ensuring that this piece of music is composed in moll and not in the parallel major. The result is a propositional sentence like the above-mentioned. Linguistic hermeneutics is a hermeneutics of context relations and pre-knowledge. Only if you know that four crosses are related to E major and C-sharp minor, you interpret the scores of the moonlight-sonata in those terms.

What is material hermeneutics in that context? As long as a piano player interprets the scores in propositional aspects, they remain silent. With the first touch and the first feeling of the feedback of the piano keys the material and bodily-sensory interpretation starts. (And it begins even earlier on the visual level, when somebody enters a room, realizes a black and white thing close to the

p. 246; Irrgang 2008a, p. 56). "Verstehen hat demnach die Form eines Lernprozesses, der durch das an ihren Folgen erkennbare Gelingen bzw. Mißlingen von Interpretationen gesteuert wird." (Jung 2002, p. 151: "Understanding has the form of a learning process that is controlled by the success and failure of interpretations, which is recognizable by the consequences." [Translated by the author, Michael Funk.])

10 On Wittgenstein´s approach in the "Philosophical Investigations" see also Funk 2010.

11 Bernhard Irrgang pays tribute to that by working out a synoptic approach including both sides of this coin in Irrgang 2007. Current paleoanthropological research supports this thesis: there is a feedback-loop of biological evolution and cultural evolution in the development of humankind since the first use of technical tools (Schrenk 2008, p. 77, p. 99, p. 122).

wall and interprets it as a tool for musical gestures.[12]) So, material interpretation is sensory interpretation. The material weight and feedback of the key involves a particular sensory perception as well as a particular human form of body movement. Thereby the piano becomes a cultural mirror that shapes our social and self-interpretations on a tacit level. While we are in interaction with the instrument and other human beings, we are involved in feedback loops and processes of implicit knowing. The body-gesture of a pianist playing the moonlight sonata is more than he is able to say. But it is a matter of knowing, related to a concrete cultural horizon and bodily tacit pre-knowledge. The cultural horizon of material hermeneutics is not only represented by the piano, but necessarily also by the traditions of piano players, who share their knowledge within social interactions. Therefore it is not necessary to read scores in an ivory tower. The shared gestures of student and teacher sitting next to each other in front of the piano are the basic point. Even if they do not say any word, within processes of trial and error, of moving the hands and copying the movements of hands, on a gestural level knowledge is shared.[13] Related to perspectivity the 1PP (the student who interacts with the piano) and the 2PP (the teacher who gives the feedback gesture) are involved in the 1PPP (the cultural horizon of musical traditions …).[14]

Perspectivity and Hermeneutics of Action

Now I would like to focus on robots and on the engineering sciences (*Technikwissenschaften*). In the current discussion, the search for an epistemic theory of the engineering sciences plays an important role (Banse et al. 2006; Irrgang 2010; Kornwachs 2012; Poser 2012, pp. 312-331). The question is related to the different structures of handcraft knowledge and scientific knowledge, the relation between sciences and the so-called "lifeworld," but also how they are interrelated within the praxis of engineering. Different approaches and methodologies are discussed, with *Technikhermeneutik* playing an important role as well. Following the impulse of Bernhard Irrgang (Irrgang 1996, pp. 56ff.), in current research on engineering sciences Hans Poser tries to overcome the defensive behavior of linguistic hermeneutics when he states:

> "Dies [die Lösung eines konkreten technischen Problems] verlangt, dass der Ingenieur zu etwas in der Lage ist, was man normalerweise als besondere Qualifikation des Geisteswissenschaftlers ansieht, nämlich eine gegebene Situation in ihrer Einzigartigkeit zu verstehen. […] So zeigt sich, dass es eine Dimension der Technikwissen-

---

12 Another example is interpreting the paper as musical scores and not as a poem in a foreign language. For visual and pictorial hermeneutics see: Fellman 1991, p. 62; Jung 2002, p. 143ff., p. 153; on visual competence see the discussions in Hug & Kriwak 2011.

13 On the current discussion about theories of knowledge-sharing and new media technologies see Sützl et al. (eds.) 2012.

14 For an approach to the epistemology of classical music, Indian Raga music(s) or Jazz see also Funk 2011; Funk 2012; Funk & Coeckelbergh 2013.

> schaften gibt, die einer Methodologie bedarf, welche bisher als reine Domäne der geisteswissenschaftlichen Hermeneutik galt." (Poser 2012, p. 327; see also Poser 2004, p. 190)

This means: An engineer needs to be able to understand a given situation in its singularity because he wants to find a concrete technical solution. Specific circumstances call for adequate technological solutions. Therefore engineers need skills that are usually associated with the "*Geisteswissenschaften*." Insofar, there is a nuance of the engineering sciences, which is shaped by a hermeneutical methodology, which has previously been related to the social sciences.

Especially the unavoidable influence of concrete given situations shapes what Thomas Rentsch discusses as "*Situationsapriori*," the apriority of situations (Rentsch 1999, pp. 68ff.; Rentsch 2003, pp. 75ff.; with respect to hermeneutics see also Joisten 2009, p. 9). And for Don Ihde, "our action, experience and knowledge is situated" (Ihde 2010, p. 41). The apriority of situations is an anthropological and existential framework of human actions, of engineering actions as well as philosophical or musical actions. Competence and knowing in handling ("*Umgehen mit*") particular constellations belongs to the 1PP. We are always part of situations. There is no isolated view from outside (3PP). And it is at this point where we can link hermeneutics and epistemology with perspectivity. "Things can only be understood from perspectivity, as the Renaissance illustrated fairly well." (Irrgang 2008a, p. 58)

What does this say about Human-Robot-Interactions? "The concept of action [...] is an ascriptive one, which is to say it is based on other people´s ascriptions to an action performed by an agent." (Janich 2012, p. 217; see also Irrgang 2005a, p. 187; Irrgang 2005b, p. 248) And those ascriptions are done from the 3PP embedded into the 1PP. In terms of Armin Grunwald: "The interpretation is made in reconstruction by external observers, or by the actor him-/herself." (Grunwald 2012, p. 192) This is related to the apriority of situations: the interpretation and understanding of situations is inherently involved in contexts of specific actions.

> "Action itself is a construct of interpretation, a phenomenon of attribution. Viewed as pure physical procedure (3PP), actions don´t differ from events (Ereignisse)." (Irrgang 2008a, p. 65)

From this point of view, a robot can "act" (Grunwald 2012, pp. 199-200), because the "objective part of technical competence-action-schemata can be implemented into machines" (Irrgang 2008a, p. 66). But the "abovementioned hypothesis is only then tenable, if the attribution is made from the perspective of an external observation, and when questions of a 'personhood' are avoided." (Grunwald 2012, p. 190) "The acting of a robot is a case of action without acting subject." (Irrgang 2008a, p. 64) Human behavior is simulated by computers in terms of the 3PP but human executions (*Vollzug*) of an acting "I" are related to the 1PP and not a matter of computer simulation or robots (Irrgang 2005a, p. 76,

p. 160; Irrgang 2005b, p. 246). Insofar, a robot could sit or stand in front of a piano and simulate a piece of music. But it could not creatively interpret the scores with personal articulation. The epistemic reason is the robot´s inability to interrelate the actions of 1PP to the mirror of the 2PP within the cultural horizon (1PPP). This cannot be substituted by the 3PP. *Technikhermeneutik* is not a way, in which robots are interpreting the world. *Technikhermeneutik* is how human beings interpret perspective actions that are always authentic constellations of 1PP, 2PP, 3PP, 1PPP, 2PPP and 3PPP. This observation is not only valid for music or philosophy, but also for engineering competence and knowledge, which is not replaceable by robots.

The 2PP and Intersubjectivity

As ascription or attributive term, viewed from the outside (3PP), robots can "act." Both interacting poles – human and robot – are seen as on the same level. When we say that the objective part of a competence-action-schema is implementable, then we focus on objective knowing, which is expressible in strict and truth-apt terms or natural laws. Now, let us ask the question about personhood and relate it to a skeptical twist. Arguing against skepticism means searching for the true foundations of objective and strict knowledge, as it has been done by René Descartes (Descartes 2009), who discovered perspectivity (Irrgang 2005a, p. 29). His basic argument against skepticism is "cogito ergo sum." He does not pay any attention to concrete other individuals (2PP). Descartes did not write: "I tell you about my thoughts and you give an intelligent response to it, therefore we are." His focus is on the acting "I" (1PP) in relation to the possibility of objective knowing (3PP).

In that classical philosophy of subjectivity the "I" (1PP) is not related to a "You" (2PP). But later in the 1840s Ludwig Feuerbach developed a concept of philosophy of *Leiblichkeit* in critical distance to Hegel. He described the importance of social interaction and the 2PP (Feuerbach 1983, §32, § 36, §§50ff.). Feuerbach´s argument was about human beings, not robots or machines, and it illustrates what is human about human knowing: the competence of social interaction related to the personal actions (1PP) and interactions with other persons (2PP).[15] We are always interacting in specific situations (*Situationsapriori*) and the hermeneutical skills of understanding and interpreting those situations characterize human knowing and acting inherently. There is no isolated *cogito* that was taken for granted by René Descartes.

According to 20th-century post-Cartesian philosophy, an adequate understanding of the 1PP is related to a "hermeneutics of the self" (Irrgang 2005b, p. 242) and the reference to another person (2PP) that is a "hermeneutics of the

15 According to paleoanthropological research, social behavior (*Sozialverhalten*) is one fundament for the development of modern humans (Schrenk 2008, p. 99). The interrelation between 1PP and 2PP belongs to that observation.

other" (Irrgang 2008a, p. 56). Human "intersubjectivity" is constituted by the interrelation between both perspectives as emphasized by Shaun Gallagher (Gallagher 1996). Paul Ricoeur is another influential author in this tradition. For him selfhood and otherness cannot be separated. We always understand who we are in social interaction with other humans. Insofar, this hermeneutics of the self is a philosophical approach that is leaving the trajectory of classical philosophy of subjectivity and *cogito = sum* (Irrgang 2005b, p. 242). Human actions are always bodily and socially constituted and thus shaped by horizons of intersubjectivity and multiperspectivity:

> "A Robot behaves within a frame. An animal behaves within a frame of behavioral patterns. But Humans act inside a horizon [that is the 1PPP, remark by Michael Funk]. For this reason, it seems to me, the problem of intersubjectivity and heterophenomenology arises against the background of the perspectivity of human subjectivity. The question regarding perspectivity of human intersubjectivity constitutes the problem of horizon for human-embodied action. Competences have horizons; they are not constituted by omnipotence." (Irrgang 2008a, p. 59)

Human actions are a matter of different points of view, of different standpoints. In social everyday life, 1PP and 2PP are the most significant perspectives. But human interaction is also embedded in cultural horizons. Language, gestures or material traditions such as handcraft technologies are examples for that. These cultural horizons constitute the platform of intersubjective understanding and communication, they shape the "we" (1PPP). And

> "[…] actions are (always, by definition) 'intentional;' moreover, anything that contains an element of action is 'intentional,' and vice versa. One of the most important forms of action is language." (Janich 2012, p. 218)

And here we can find the link to social interaction and the interrelation between 1PP and 2PP:

> "Language comprehension and generation […] remains tied to intentionally acting human agents, even in forms of communication that remain invariant regardless of the speakers and listeners. […] a translation must capture what it is that one person wants of another when they speak to them. (And this is not decided by an observer located at an Archimedean vantage point outside of the communicative situation. In case of doubt, the two interlocutors themselves must reach a consensus on it.)" (Janich 2012, p. 228)

Material hermeneutics is related to *Technikhermeneutik* as we have seen with respect to perspectivity and processes of knowing. Especially the interactions between humans or humans and robots are characterized by body language, gestures etc. And "since languages and the cultures in which these languages are embedded differ, this approach naturally suggests paying attention to cultural differences" (Coeckelbergh 2011a, p. 65). If we talk about "material hermeneutics" or "*Technikhermeneutik*" related to cultural perspective praxis, what does "material" mean? What is material culture?

Material Culture and Embodied (Leibliche) Technics

> "Material culture may be defined as the human significance of the totality of tangible artifacts that humans have produced. These artifacts range from the mundane and perishable to the monumental and enduring, and have been linked together in distinctive ways across place and time." (Borgmann 2005, p. 1172)

Borgmann emphasizes the developments since the industrial revolution till today (Borgmann 2005, pp. 1173ff.). One important aspect can be seen in the interrelations between technologies and the rise of modern natural sciences:

> "Research and development have to this day been the major sources of productivity growth and thus of an exploding material culture. By now technology and science have so fulsomely embraced one another that it has become fashionable to see them as one creature – technoscience (Ihde and Selinger 2003)." (Borgmann 2005, p. 1173)

Our knowledge about material culture is linked to *leibliches*/embodied knowing: sensorimotor movements (form 1) and the associated perceptions (form 2), emotions (form 3), propositions (form 4) and philosophical-reflections (form 5). There is no Cartesian dualism (between *res extensa*: the body, and *res cogitans*: the mind) in human praxis. Of course we can use the words "*res extensa*" and "*res cogitans*" to illustrate something, like I am using the words "form 1" or "form 2" in this text. But it is important to see that this is always done by a philosophical scalpel, with which we are cutting the things that happen into systematic pieces. Insofar it is not wrong to separate words like "materiality," "culture," "technology" or "sciences." But on the other hand, it is also important to focus on the cut surface. This is done with terms like "material hermeneutics," "*Technikhermeneutik*," "material culture" or "technoscience." And here we can realize that technical-material frameworks have shaped human cultures at least since the first use of hand axes (Müller-Beck 2008, pp. 18ff., p. 28, p. 30; see also Irrgang 2008b, pp. 55ff.). Insofar the term "material culture" is not only reserved for technological developments since the industrial revolution or for "technoscience." Understanding the meaning of material culture implies understanding the interrelations between cultural evolution and biological-genetic evolution. There is also a cut surface between nature/biology and culture. A genuine feedback between biological evolution and cultural evolution caused the rise of modern *homo sapiens* on the basis of handcraft-tool-culture (Schrenk 2008, p. 77, p. 99, p. 122). Within cultural praxis we are shaping our genes. One example is the capacity for digesting lactose that has been developed in early livestock breeding cultures (Leonardi et al. 2012). There is a long tradition of tacit and implicit breeding knowledge, which precedes written records and is an important part of our material cultural traditions. My hypothesis is: We can understand Human-Robot-Interactions as new forms of those traditions, where human cultural and human biological evolutions are intertwined. Robots are pieces of material culture, which are to be embedded in frameworks and horizons of perspectivity, related to the five forms of knowledge.

> "[...] how we interact with robots and the precise form of our relation with them partly depends on our own character, identity, and personal history. [...] The historical aspect is often neglected in reflections on human-robot relations." (Coeckelbergh 2011b, p. 201)

This is what Michael Polanyi describes as "personal knowledge" (Polanyi 2002). The individual perceptual, emotional and sensorimotor histories of human beings carry personal knowing. Within our histories we not only remember visual perceptions or information, but also body movements, which are the basis for sensory embodied knowing (form 1 and 2)[16] and structures of emotional (form 3), propositional (form 4) and philosophical-reflexive (form 5) remembering are cross-linked. Individual processes are always mirrored on a perspectival prism: social interactions (2PP) within a cultural horizon (1PPP), which carries theoretical and objective knowing (3PP). But as we have seen, especially with robots we are enhancing the framework of intersubjectivity, the "quasi other" (Ihde and Coeckelbergh) that is the 2PP.

A basis for material-technical handcraft cultures in general is the human competence to remember long sequences of body movements related to particular perceptive, social and material feedback. That is the fundament of human handcraft cultures since the first use of hand axes (Irrgang 2008b, p. 55; Irrgang 2009a, pp. 47ff., pp. 68ff., pp. 91ff., pp. 109ff.; Irrgang 2009b, pp. 7ff.). This epistemological approach does not deny or diminish the importance of semantic or propositional information. But an explanation of human culture has to start from the aspect of technical praxis; and insofar a philosophical approach of *Technikhermeneutik* is on one level with the state of the art in paleoanthropology or paleogenetics (Irrgang 2009a, pp. 47-82). Here the transdisciplinary gap is bridged. Within processes of robotics-research the approach of *Technikhermeneutik* provides a contribution to the development and framework of transdisciplinary oriented Philosophy of Technology and Technology Assessment.

Robots can be seen as another technical tool,[17] thus another piece of the development of human material culture that is involved in the feedback loop between biological and cultural evolution.

---

16 For studies of motor body movements in the arts see De Preester (ed.) 2013.

17 "Roboter, die 'autonom' handeln können, [...] bleiben Werkzeuge und werden nicht zu handelnden Subjekten." (Irrgang 2005a, p. 165: "Robots that are able to act autonomously [...] remain tools and do not become acting subjects." [Translated by the author, Michael Funk.]) "Weder der Mensch noch die Realität sollten durch Technik ersetzt werden, sondern vielmehr ihr Werkzeug- und Mittelcharakter betont werden, um die Ziele ihres Einsatzes besser reflektieren zu können." (Irrgang 2005a, p. 212: "Neither humans nor reality should be replaced by technics. Instead we should put more emphasis on its status as a tool or instrument in order to be able to better reflect the purposes of its use." [Trans-

> "Deshalb wird hier vorgeschlagen, Robotik ausschließlich als Werkzeug bzw. Mittel für menschliche Zwecke aufzufassen – zumal der gegenwärtige Stand der technischen Entwicklung kaum eine andere Deutung zulässt. [...] Unabhängig davon ist generell Versuchen zu widersprechen, spekulativ die Kluft zwischen menschlicher und künstlicher Intelligenz einzuebnen oder deren Einebnung als kulturelles Ziel auszugeben." (Christaller et al. 2001, p. 127:
>
> "Hence it is suggested that robotics is understood only as tool or means for human purposes – especially as the current technological state of the art allows no other interpretation. [...] Anyway, attempts at bridging the gap between human and artificial intelligence or at establishing this as a cultural target are to be rejected." [Translated by the author, Michael Funk.])

What does this mean for robotics in German and Japan culture?

Robotics in Germany and Japan

Technologytransfer between the Western World and East Asia has a long tradition. The Silk Road is an example for that. With technologies we always share perceptions, e.g. with spices we share a specific taste or with musical instruments a certain sound. And there is a genuine history of technology transfer between Germany and Japan as well, starting in the middle of the 19th-century and continuing till today.[18] The mid-19th-century developments in Japan caused an opening to European and Western influences related to what we are used to call "industrialization" or "modernization." The synthesis between European industrialized technologies and East Asian traditions mostly led to forms of so-called "alternative modernizations." Alternative modernization means understanding and handling modern technologies, in non-Western cultures. Sharing technologies is more than sharing artifacts or physical and mathematical information. Moreover we also share values, tacit knowing and other factors of cultural embedding (Irrgang 2006). Saying it in words of perspectivity: there is not "the one and only" modernity, but many cultural horizons of handling and understanding modern technolog*ies*. If we talk about "alternative modernity" we talk through the 1PPP. If something is "alternative" it is only "alternative" viewed through the scope or glasses of a native cultural horizon. The slogan "alternative modernity" belongs to the 2PPP, the focus of other cultural horizons.

Related to the development of robotics, the engineering sciences and the research on robots, we can see a scientific community that tries to establish transcultural knowledge that should be valid on a physical and mathematical level for robots all over the world (3PPP). The WWW, online journals and international robotics-conferences are amplifying that effect: the expertise in developing, constructing and repairing robots becomes globalized like the problem of waste disposal. On the other hand, the expertise – the perspective interrelation of the five

---

lated by the author, Michael Funk.])

18 For further studies and detailed information see Pauer (ed.) 1992.

forms of knowledge related to concrete humans using robots – is strongly shaped by the 1PPP. The way we handle robots is not globalized. And it is at this point where we find the most obvious differences between robotics in Germany and Japan.

> "Different cultures have different views on autonomy and human dignity." (Carpurro 2009, p. 122)
>
> Thus "Robots are a mirror of shared cultural values that show to us and to others who we want to be. We redefine ourselves in comparison with robots in a similar way as we redefine ourselves in comparison with animals or gods." (Carpurro 2009, p. 120)
>
> "For instance, there seem to be differences between Europe and Japan in the way robots are perceived: the 'slave' model seems to be more dominant in Europe whereas Japan seems to go for the 'companion' model." (Coeckelbergh 2001b, p. 201)

One prominent example is the religious background. As a consequence of Christian influence, Europeans are used to understand humans as "*Imago Dei*," as beings with a personality. Robots just like stones or trees have no soul. But for example in Japanese Shintoism the animistic background allows seeing and treating robots as beings with an incorporated "soul." Another point is related to the image of robots in popular culture.[19] But these cultural effects are also not separable from economic developments[20]:

> "The differences on the surface, such as more enthusiasm in investing in advanced robot development and popularity in mass media, might result from social demands in order to cope with the lack of young workers due to the aged society, rather than from the cultural and religious background. If Japanese people are deemed to take more positive and less ethical attitudes towards robots, this might result from subtle differences in conceptualization concerning robots as artificial humans and the shared belief about the Japanese historical background of animism and polytheism." (Nishida 2009, p. 112)

The concrete situations (*Situationsapriori*) in which people interact with robots are involved into a cultural horizon (1PPP), in which the 2PP is like a mirror of those subtle differences. These situations cannot be substituted by the 3PPP, the horizon of "objective," physical and mathematical knowing. Insofar, there is no difference between robots and hand axes: physical theory is the same all over the world, but the ways in which humans handle those tools within social and cul-

---

19 See also the chapter "Robots in Japanese Popular Culture" (Maika Nakao) in this book.

20 "The acceptance of technologies and thus their demand may be higher in technophile economies (Japan is generally considered as being one of them) than in more conservative ones." (Decker et al. 2011, p. 39) "Japaner arbeiten mit Robotern, andere Nationen mit billigen Immigranten. Die Japaner werden davon langfristig einen enormen Vorteil haben." (Irrgang 2005a, p. 152: "Japanese people are working with robots, other nations with cheap immigrants. In the long run, this will cause enormous benefits for Japan." [Translated by the author, Michael Funk.])

tural frameworks are not congruent. From the perspective of modern sciences (3PP and 3PPP), it may be considered as irrational to treat a robot like a personal being with a soul. For certain users (1PP) within an animistic religious background (1PPP), this is not irrational, it is taken for granted.

Conclusion

In this approach, a pragmatic hermeneutics of perspectivity has been developed. This is not a linguistic hermeneutics related to the interpretation of texts but a philosophy of multi-perspective knowing and understanding in terms of material culture. The basic point is to demonstrate the relations between human processes of knowing and the way we treat robots as technical tools. Robots will enhance and shape but not replace human knowing. Insofar robots can be seen as tools just like hand axes or cars. But on the other hand, of course, robots are more complex technical systems that can "act" (as far as we accept "action" as an ascription from the 3PP). Robots cannot replace the human perceptive and emotional, moreover not the philosophical-reflexive, forms of knowledge. Anticipating the replacement of humans by robots is far away from the technical state of the art and bears no useful cultural values. But robots are also a challenge for humans to develop new forms of interrelating the perspectives (1PP, 2PP and 3PP) within cultural horizons (1PPP, 2PPP and 3PPP). Especially the aspect of verbal and gestural communication is relevant for the impact of the 2PP. As far as we know, a hand axe does not respond on a verbal level to what the user does. But (humanoid service) robots may do so. They appear as "quasi others" as a mirror that is not congruent but close to the human 2PP and a challenge for humans to learn new competences.

> "A philosophical hermeneutics is required as a rational (scientific) method by which for instance two competing interpretations (i.e. translations) of the same text can be judged by explicit criteria. To date, no such science exists." (Janich 2012, p. 228)

In terms of perspectivity and a hermeneutical epistemology such a judgment, for example within human-robot verbal interactions, is only possible from the 3PP within a framework of the 3PPP. Human social interactions are mostly driven by non-explicit tacit knowing (1PP and 2PP). But what we can do is to develop a cultural, social and institutional framework within the 1PPP and 3PPP that carries and shares explicit forms of knowing. We need to develop cultural frameworks that allow evolving the competences for using robots in everyday life and thinking about the related social and ethical impacts; namely Philosophy of Technology (Irrgang 2008b, pp. 341ff.), as well as Technology Assessment (Grunwald 2010).

> „Wir sollten nicht das Vermögen der Roboter fürchten, sondern unser eigenes technisches Unvermögen.“ (Irrgang 2005a, p. 161: "We should not fear the capabilities of robots, but our own technical inability." [Translated by the author, Michael Funk])

References

Achterhuis, H. (ed.) 2001: American Philosophy of Technology: The Empirical Turn, Bloomington a.o.

Banse, G., A. Grunwald, W. König & G. Ropohl (eds.) 2006: Erkennen und Gestalten. Eine Theorie der Technikwissenschaften, Berlin.

Borgmann, A. 2005: "Material Culture." In: Mitcham, C. (ed.) 2005: Encyclopedia of Science, Technology, and Ethics, Detroit a.o., pp. 1172-1176.

Carpurro, R. 2009: "Ethics and Robotics." In: Capurro, R. & M. Nagenborg (eds.) 2009: Ethics and Robotics, Heidelberg, pp. 117-123.

Christaller, Th. 2007: "Vorwort." In: Kajita, Sh. (ed.) 2007: Humanoide Roboter. Theorie und Technik des Künstlichen Menschen, Berlin, p. ix.

Christaller, T., M. Decker, J.-M. Gilsbach, G. Hirzinger, K. Lauterbach, E. Schweighofer, G. Schweitzer & D. Sturma 2001: Robotik. Perspektiven für menschliches Handeln in der zukünftigen Gesellschaft, Berlin u.a.

Coeckelbergh, M. 2011a: "You, Robot: On the Linguistic Construction of Artificial Others." In: AI & Soc (2011), 26, pp. 61-69.

Coeckelbergh, M. 2011b: "Humans, Animals, and Robots: A phenomenological Approach to Human-Robot Relations." In: Int J Soc Robot (2011), 3, pp. 197-204.

Decker, M. 2010: "Ein Abbild des Menschen: Humanoide Roboter." In: Bölker, M., M. Gutmann & W. Hesse (eds.): Information und Menschenbild, Berlin & Heidelberg, pp. 41-62.

Decker, M. 2011: "Serviceroboter in medizinischen Anwendungen. Eine interdisziplinäre Problemstellung." In: Maring, M. (ed.) 2011: Fallstudien zur Ethik in Wissenschaft, Wirtschaft, Technik und Gesellschaft, Karlsruhe, pp. 249-255.

Decker, M., R. Dillmann, Th. Dreier, M. Fischer, M. Gutmann, I. Ott & I. Spiecker (known as Döhmann) 2011: "Service Robotics: Do You Know Your New Companion? Framing an Interdisciplinary Technology Assessment." In: Poiesis Prax, 2011 November, 8 (1), pp. 25-44.

Descartes, R. 2009: Meditationen. Mit sämtlichen Einwänden und Erwiderungen, Hamburg.

De Preester, H. (ed.) 2013: Moving Imagination – Explorations of Gesture and Inner Movement in the Arts, Amsterdam & Philadelphia.

Dreyfus, H. 1979. What Computers Can`t Do. Revised Edition, New York.

Fellman, F. 1991: Symbolischer Pragmatismus. Hermeneutik nach Dilthey, Reinbek.

Feuerbach, L. 1983: Grundsätze der Philosophie der Zukunft. Kritische Ausgabe mit Einleitung und Anmerkungen von Gerhart Schmidt. 3. Auflage, Frankfurt a.M.

Funk, M. 2010: "Verstehen und Wissen. Ludwig Wittgensteins Philosophie der Technik." In: Leidl, L. & D. Pinzer (eds.) 2010: Technikhermeneutik. Technik zwischen Verstehen und Gestalten. Frankfurt a.M. a.o., pp. 75-88.

Funk, M. 2011: "Vom Blick zum Klang – Was wissen die Noten über mein Klavierspiel?" In: Hug, Th. & A. Kriwak (eds.) 2011: Visuelle Kompetenz. Beiträge des interfakultären Forums Innsbruck Media Studies, Innsbruck, pp. 270-284.

Funk, M. 2012: "Leibbilden, Notenbilden, Onlinebilden - Wie wissen wir Musik zu teilen?" In: Sützl, W., F. Stalder, R. Maier & Th. Hug (eds.) 2012: Media, Knowledge and Education: Cultures and Ethics of Sharing. Medien – Wissen – Bildung: Kulturen und Ethiken des Teilens, Innsbruck, pp. 209-226.

Funk, M. & M. Coeckelbergh 2013: "Is Gesture Knowledge? A Philosophical Approach to the Epistemology of Musical Gestures." In: De Preester, H. (ed.) 2013: Moving Imagination – Explorations of Gesture and Inner Movement in the Arts, Amsterdam & Philadelphia, pp. 113-132.
Gallagher, Sh. 1996: "The Moral Significance of Primitive Self-Consciousness." In: Ethics 107, pp. 129-140.
Gill, S. P. 2008: "Socio-Ethics of Interaction with Intelligent Interactive Technologies." In: AI & Society, 22(3), pp. 283-300.
Grunwald, A. 2010: Technikfolgenabschätzung – eine Einführung. Zweite, grundlegend überarbeitete und wesentlich erweiterte Auflage, Berlin.
Grunwald, A. 2012: "Can Robots Plan, and What Does the Answer to this Question Mean?" In: Decker, M. & M. Gutmann (eds.) 2012: Robo- and Informationethics. Some Fundamentals. Hermeneutics and Anthropology Vol. 3, Berlin a.o., pp. 189-209.
Heidegger, M. 2006: Sein und Zeit. 19. A., Tübingen.
Hirukawa, H. 2007: "Kapitel 1. Überblick Humanoide Roboter." In: Kajita, Sh. (ed.) 2007: Humanoide Roboter. Theorie und Technik des Künstlichen Menschen, Berlin, pp. 1-14.
Hug, Th. & A. Kriwak (eds.) 2011: Visuelle Kompetenz. Beiträge des interfakultären Forums Innsbruck Media Studies, Innsbruck.
Husserl, E. 2012: Die Krisis der europäischen Wissenschaften und die transzendentale Phänomenologie, Hamburg.
Ihde, D. 1990: Technology and the Lifeworld. From Garden to Earth, Bloomington.
Ihde, D. 1993: Postphenomenology. Essays in the Postmodern Context, Evanston.
Ihde, D. 1998: Expanding Hermeneutics. Visualism in Science, Evanston.
Ihde, D. 2010: "Phenomenologists and Robots." In: Ihde, D. 2010: Embodied Technics. Copenhagen, pp. 37-53.
Ihde, D. & E. Selinger (eds.) 2003: Chasing Technoscience. Matrix for Materiality, Bloomington.
Irrgang, B. 1996: "Von der Technologiefolgenabschätzung zur Technologiegestaltung. Plädoyer für eine Technikhermeneutik." In: Jahrbuch für christliche Sozialwissenschaften, 37 (1996), pp. 51-66.
Irrgang, B. 2005a: Posthumanes Menschsein? Künstliche Intelligenz, Cyberspace, Roboter, Cyborgs und Designer-Menschen – Anthropologie des künstlichen Menschen im 21. Jahrhundert, Stuttgart.
Irrgang, B. 2005b: "Ethical Acts (Action) in Robotics." In: Brey, Ph., F. Grodzinsky & L. Introna (eds.) 2005: Ethics of New Information Technology. Proceedings of the Sixth International Conference of Computer Ethics: Philosophical Enquiry (CEPE2005). July 17-19. Enschede. The Netherlands, Center for Telematics and Information Technology (CTIT), Enschede, pp. 241-250.
Irrgang, B. 2006: Technologietransfer transkulturell. Komparative Hermeneutik von Technik in Europa, Indien und China. Dresden Philosophy of Technology Studies 1, Frankfurt a.M. a.o.
Irrgang, B. 2008a: "Intersubjectivity, 'Other Intelligences' and the Philosophical Constitution of the Human-Robotics-Interaction." In: Prajna Vihara, Vol. 9, Nr. 2, July-December, 2008, pp. 56-69.
Irrgang, B. 2008b: Philosophie der Technik, Darmstadt.
Irrgang, B. 2009a: Der Leib des Menschen. Grundriss einer phänomenologisch-hermeneutischen Anthropologie, Stuttgart.

Irrgang, B. 2009b: Grundriss der Technikphilosophie. Hermeneutisch-phänomenologische Perspektiven, Würzburg.

Irrgang, B. 2010: Von der technischen Konstruktion zum technologischen Design. Philosophische Versuche zur Theorie der Ingenieurpraxis, Berlin.

Janich, P. 1999: "Substitution kommunikativer Kompetenz?" In: Decker, M. (ed.) 1999: Robotik. Einführung in eine interdisziplinäre Diskussion, Bad Neuenahr-Ahrweiler, pp. 17-31.

Janich, P. 2012: "Between Innovative Forms of Technology and Human Autonomy: Possibilities and Limitations of the Technical Substitution of Human Work." In: Decker, M. & M. Gutmann (eds.) 2012: Robo- and Informationethics. Some Fundamentals. Hermeneutics and Anthropology Vol. 3, Berlin a.o., pp. 211-230.

Johnson, M. 1987: The Body in the Mind: The Bodily Basis of Imagination, Reason, and Meaning, Chicago.

Joisten, K. 2009: Philosophische Hermeneutik, Berlin.

Jung, M. 2002: Hermeneutik zur Einführung. 2. A., Hamburg.

Kajita, Sh. 2007: "Vorwort." In: Kajita, Sh. (ed.) 2007: Humanoide Roboter. Theorie und Technik des Künstlichen Menschen, Berlin, p. xi.

Kemp, Ch. S. et al. 2008: "Humanoids." In: Sicilliano, B. & P. Khatib 2008: Springer Handbook of Robotics, Berlin & Heidelberg, pp. 1309-1333.

Knoll, A. & Th. Christaller 2003: Robotik, Frankfurt a.M.

Kornwachs, K. 2012: Strukturen technologischen Wissens. Analytische Studien zu einer Wissenschaftstheorie der Technik, Berlin.

Kurthen, M. 1994: Hermeneutische Kognitionswissenschaft. Die Krise der Orthodoxie, Bonn.

Leonardi M., P. Gerbault, M. G. Thomas & J. Burger 2012: "The Evolution of Lactase Persistence in Europe. A Synthesis of Archaeological and Genetic Evidence." In: International Dairy Journal 22, 2012, pp. 88-97

Mahrenholz, S. 2011: Kreativität. Eine philosophische Analyse, Berlin.

Müller-Beck, H. 2008: Die Steinzeit. Der Weg des Menschen in die Geschichte. 4. A., München.

Nishida, T. 2009: "Towards Robots with Good Will." In: Capurro, R. & M. Nagenborg (eds.) 2009: Ethics and Robotics, Heidelberg, pp. 105-116.

Noë, A. 2004: Action in Perception. Cambrigde.

Pauer, Erich, (ed.) 1992: Technologietransfer Deutschland – Japan von 1850 bis zur Gegenwart. Monographien aus dem Deutschen Institut für Japanstudien der Philipp-Franz-von-Siebold-Stiftung. Band 2, München.

Piaget, J. 1954: The Construction of Reality in the Child, New York.

Polanyi, M. 1969: Knowing and Being. Essays by Michael Polanyi. Ed. by Marjorie Grene, Chicago.

Polanyi, M. 2002: Personal knowledge. Towards a post-critical philosophy, London a.o.

Polanyi, M. 2009: The tacit Dimension. With a new foreword by Amartya Sen, Chicago & London: The University of Chicago Press.

Poser, H. 2012: Wissenschaftstheorie. Eine philosophische Einführung, Stuttgart.

Poser, H. 2004: "Technikwissenschaften im Kontext der Wissenschaften." In: Banse, G. & G. Ropohl (eds.) 2004: Wissenskonzepte für die Ingenieurpraxis. Technikwissenschaften zwischen Erkennen und Handeln. VDI-Report 35, Düsseldorf, pp. 175-193.

Rentsch, Th. 1999: Die Konstitution der Moralität. Transzendentale Anthropologie und praktische Philosophie, Frankfurt a.M.

Rentsch, Th. 2003: Heidegger und Wittgenstein. Existenzial- und Sprachanalysen zu den Grundlagen philosophischer Anthropologie, Stuttgart.

Ryle, G. 1946: "Knowing How and Knowing That." In: Proc. Aristot. Soc. 46 (1945/46), pp. 1-16.

Schrenk, F. 2008: Die Frühzeit des Menschen. Der Weg zum Homo sapiens. 5. A., München.

Sützl, W., F. Stalder, R. Maier & Th. Hug (eds.) 2012: Media, Knowledge and Education: Cultures and Ethics of Sharing. Medien – Wissen – Bildung: Kulturen und Ethiken des Teilens, Innsbruck.

Varela, F. J., E. Thompson & E. Rosch 1997: The Embodied Mind. Cognitive Science and Human Experience. Sixth Printing, Cambridge & London.

Verbeek, P.-P. 2005: What Things Do. Philosophical Reflections on Technology, Agency, and Design, University Park Pa.

Wittgenstein, L. 2006: "Philosophische Untersuchungen." In: Wittgenstein, L. 2006: Werkausgabe Band 1, Frankfurt a.M., pp. 225-577.

Technology Assessment

Who is taking over?
Technology Assessment of Autonomous (Service) Robots

Michael Decker

Introduction

Several recent events put robot systems into public attention. In connection with the Fukushima nuclear disaster (Stöcker 2011) and the explosion of the oil rig "Deepwater Horizon" the use of robot systems to combat these catastrophes was discussed. During the oil spill, robots were used to close the leak. According to media reports, an underwater robot was used to cut off a torn pipe and to lower a containment dome over the leak (Spiegelonline 04.06.2010). Ariel Bleicher (2010) drew "the first Gulf Spill′s lessons for Robotics. The demands on the largest underwater robotics armada ever fielded show that ROVs need better automation" (while ROV means remotely operated vehicles).[1] "To help eliminate human error, ROV manufacturers like Schilling Robotics are developing computer software to automate some of the standard things ROVs do." This automation should not only reduce the time a robot needs to complete a task, but also improve the quality of its results. Moreover, robotics is also used for the investigation of subsea hydrocarbon plums caused by the blowout. An international group from the Australian Centre for Field Robotics at the University of Sydney and from the Deep Submergence Laboratory at Woods Hole Oceanographic Institution employed conventional oceanographic sampling techniques along with the SENTRY autonomous underwater vehicle to confirm the existence of a coherent subsea hydrocarbon plume, to map the plume′s spatial extent and finally to collect targeted water samples from within the plume for later laboratory analysis (Jakuba et al. 2010).

Robot systems for deep-sea operation belong to the so-called "expansion robots" (Christaller et al. 2001, p. 217). They expand the people′s scope of action and enable them to overcome barriers and be "remotely present," i.e. to act at places they cannot access directly. This inaccessibility can result from large distances (like in space), ratios (like in the micro- or nanometre range), or physical barriers. Remote presence concepts can be used in minimally invasive surgery to transmit the hand movements of the surgeon intuitively, controllably, and commensurately to the instruments applied. Danger for the human operator may also constitute a barrier which remote presence may help to overcome – apart from deep-sea applications this could also include the disposal of explosive materials, inspection or dismantling of nuclear power stations, medical radiation, etc. In the framework of an interdisciplinary technology assessment

1 For further information see also Bleicher 2011.

these applications were classified as innovations which should be "comprehensively promoted."

With regard to the distinction between "industrial robots" and "service robots," these expansion robots can be described as intermediates. Unlike conventional industrial robots, they are not used in factory buildings where the complete environment is adapted to the use of robot systems. In most of the cases industrial robots are even operated in a safety cage to keep humans off the working space of the robot. But nevertheless, they are also not employed in our everyday environment close to human beings to perform their tasks, which could be expected for most services. Even though expansion robots are used outside factory buildings, this normally takes place in a professional context. The operators of the robots which were used during the oil spill were experts which are familiar with both the problem of oil leakages and the use of robots in this field. "Third parties" will neither be encountered in the deep sea nor in space nor in an operating room. Only a limited group of people gets in contact with the robot. All these people can be trained, if necessary even take an exam, to be able to operate the robot system and comply with the respective safety regulations.

A closer look at the areas of application of today´s service robot systems reveals that out of the 63,000 service robots for commercial applications sold worldwide until the end of 2008 the greater number were used in the field of defence, rescue, and security (30%), followed by agriculture (23%), especially milking robots (World Robotics 2009). These are areas where the service robots – in the sense of expansion robots – are operated and supervised by a human expert and/or operated in a protected surrounding. Most services, however, are characterized by the fact that they have to be performed in an environment full of people (one example might be the cleaning of train stations) or directly involve a human being (museum guide, nursing or elderly care). The people in contact with these robots can only be trained to a limited extent as robotics experts. Thus these services implicate that a robotic layperson can and has to interact with robots and that third parties will encounter a robot´s direct environment. Furthermore, these services are performed in everyday life, which can only be adapted to a limited extent to the employment of robots. This combination entails grand challenges, both for the technical realization of service robots and the societal environment where they are employed. Humanoid robots play a special role in the context of these services which directly involve a human being.

Robot systems are normally used as a means to an end. This is underlined by the case studies on expansion robots. To expand the people´s scope of action "across barriers," the robot has to be made able to complete the desired task beyond this barrier. The "end" is the sealing of the oil leakage. Therefore the pipe has to be cut off and the containment dome has to be put over the leak. A human being would not be able to work in this depth of the sea without elaborate protective equipment. If a robot exists which can be used as a means to this end, then it replaces the human being as actor in this context. Regarding the application of robot systems, one can therefore speak of "a replaceability of the human being" (Decker 2000) in specific action contexts.

In the following, this replaceability shall be discussed from different scientific disciplinary perspectives. I will start with a short description of the current definition of service robots. This will be followed by an excursus on humanoid robots. According to some researchers, they play a special role in connection with services directly involving human beings since the intuitive interaction with robots can be supported by their humanoid shape (Behnke 2008). To sum up, I will present some considerations from the perspective of technology assessment.

Multisdisciplinary Questions on TA of Service Robots

"Service robots" are often indirectly defined as "non-production robots." The International Federation of Robotics (IFR) states on its website:

> "Service robots have no strict internationally accepted definition, which, among other things, delimits them from other types of equipment, in particular the manipulating industrial robot. IFR, however, have adopted a preliminary definition:
>
> A service robot is a robot which operates semi- or fully autonomously to perform services useful to the well-being of humans and equipment, excluding manufacturing operations." (IFR)

The Fraunhofer Institute for Manufacturing Engineering and Automation (Fraunhofer IPA) phrased it in a similar way:

> "A service robot is a freely programmable mobile device carrying out services either partially or fully automatically. Services are activities that do not contribute to the direct industrial manufacture of goods, but to the performance of services for humans and institutions." (Schraft et al 2004, p. 9)

The definition by Engelhardt and Edwards refers to descriptions like "sense," "think" and "act" and reads as follows:

> "[…] systems that function as smart, programmable tools, that can sense, think, and act to benefit or enable humans or extend/enhance human productivity." (Quoted after Hüttenrauch 2006, p. 3.)

Services in the private sector are a particular challenge for service robotics since both the environment and the user of the robot can vary considerably. Therefore robot systems for private use shall be taken as preferred case studies here. Robotics provides applications for all age groups: Toy robots, entertainment robots, kitchen aids, assistant robots, care robots for elder and sick people, etc. The service sector makes high demands on robots.

They have to be able to move in an unknown environment (private flat) that is not geared to them and perform a number of different tasks. If the programming efforts prior to the start of operation should be still acceptable for robotic laypersons, most of the adaptation to the new environment and the new user has to be done by the robot system itself. This technical problem is even more critical for older users and those in need of care, since they are often cognitively unable to give the necessary instructions. In the following we will outline the questions that are relevant for a technology assessment of service robots for individual disciplines and discuss the technological, economic, legal, ethical, and psychological replaceability.[2]

Technological Perspective

The successful provision of a service is already a big technological challenge. This can be compared with a "checklist" which can be compiled for a particular service. The service "vacuum cleaning" is provided successfully if the floor is clean, and this is done without damaging furniture, without making too much noise, within a reasonable time, etc. If the vacuum cleaning robot has met these requirements, the service is – in technical terms – performed successfully. A basic requirement in the private environment is that the robot has to be able to find its way "autonomously" in a surrounding which has to date been unknown and that it can adapt to the environment in which it has to perform its service. To put it briefly: The robot has to be enabled to learn. Here we take different scientific approaches, which aim, among others, at learning "like human beings" ("learning like a child" ["Xpero Project"][3], "learning by demonstrating" ["ARMAR Project"], etc.) where "trial and error" play a central role. A humanoid stature (torso, head, arms and legs, ...) is often considered to be an advantage for learning. On the one hand it animates people to interact with the robot, on the other hand the robot is "physically" adapted to an environment which is optimized for human beings (steps adjusted to the length of human legs, doorways, signs at "eye level," etc. (see below) (Behnke 2008, p. 5). While concerning the last aspect "humanoid" just means having human dimensions and movement abilities, making the robot even more manlike can be an interesting aspect to support learning. Then we would be speaking of "android" or "gynoid" robots with a "confusingly similar" appearance to human beings. This "being like humans" could become relevant when it comes to the technical realization of so-called "soft skills" like friendliness, helpfulness, etc. which are related to the provision of services (see following paragraph).

Economic Perspective

From an economic perspective, services are first of all special "goods." Services are predominantly immaterial. They are experience goods, their quality can only be assessed once they are actually used by the customer(s). Due to the human interaction during the performance of the service the possibilities for standardiza-

2 This multi-disciplinary perspective was developed in Decker et al. 2011

3 Press release: http://www.xpero.org/portal/readarticle.php?article_id=11 (accessed Mai 30, 2011)

tion are rather limited. However, standardized processes are an indispensable precondition for the use of robots. The simultaneousness of production and consumption and the consequential direct relation between service provider and customer are the reason why services cannot be stored, exchanged, or sold again. Original innovations in the service sector are not patentable and at the same time easy to imitate. This makes it difficult to amortize the innovation costs completely and is therefore an innovation barrier. The introduction of service robots is accompanied by numerous questions, especially concerning standardization and patenting of services.

From a microeconomic perspective the following aspects are relevant and integrated into a comprehensive cost-benefit analysis: What is the motivation for individual actors to develop or use service robots (e.g. lack of nursing staff in an "aging society" and the resulting profit opportunities)? Which costs incur during the development of the robots (technical and non-technical costs)? Which costs incur during the use of the robots (availability of qualified personnel who operates the robot and is "complementary;" costs for the modification of public places; costs for creating acceptance)? etc. In this context further questions arise. Who bears which costs? Which revenues accrue? Who receives them? Are those who bear the costs also the ones who profit from the revenues? If this is not the case, we can speak of market failure, i.e., the performance of the market yields a result that is not ideal from the societal point of view. This could call for economic-political action.

From a macroeconomic perspective it is necessary to assess the importance of the service sector for the national economy. This can be done using the classification of economic activities or the ISIC (International Standard Industry Classification) which assigns specific activities to specific areas. The service sector includes numerous activities; of course their potential for employing robots has to be evaluated individually. Furthermore, it is important to identify the relevant markets. The acceptance of technologies and thus their demand may be higher in technophilic economies (Japan is generally considered as being one of them) than in more conservative ones. An aggregated analysis of the effects for the labour markets, for example, does not only consider those jobs which might be replaced by robots but also includes especially those which are newly created in the course of innovation.

Legal Perspective

Depending on the field where service robots are used, different legal questions arise. We can distinguish between those concerning the relation citizen-citizen (civil law) and others concerning the relation between the state and the citizen (public law). Being a regulatory law, public law has to restrict economic activities which collide, e.g., with the rights of others or the common good. Here one major problem are governmental decisions under uncertainty. If and how the legislative authority intervenes depends on prognostic assumptions whose future fulfilment is uncertain. It is not foreseeable if service robots will cause considerable damage to people and objects. It is also unclear if these are (normal) requirements for production safety which are already covered by the existing legal foundations of private liability law. Or do we have to assume a generally dangerous activity – in line with the liability regulations of genetic engineering – which would require absolute liability? We also have to consider secondary objectives of governmental actions: The promotion of innovation can only be successful if the different liability scenarios do not regulate the entrepreneurial (and private) development in such a strict way that further developments do not pay off. The different scenarios described here are always guiding for the different legal analyses and assessments. Social law for example, which is especially relevant for services in the field of care (age, disability, sickness), includes a number of special requirements, some of them induced by constitutional law. They differ significantly from the legal framework conditions for service robots in, e.g., agriculture.

From the perspective of civil law, where the relation citizen-citizen is in the focus of legal considerations, it is mainly a question of liability of those who plan, manufacture, sell, and finally use service robots to the integrity of legally protected goods of those people who get in contact with service robots. Here the existing regulation instruments should be made applicable to the new problems of warranty and hazard. This refers to the drafting of contracts, especially regarding the risk allocation in the General Terms and Conditions as well as general questions of liability for damages to third parties. The formulation of due diligence and liability standards is a central element here. If the requirements are too strict, this will impede – or even prevent – the manufacturing, distribution, and use of service robots; if the requirements are too low, the use is seen with even more scepticism the more defect-prone the relevant service robots turn out to be. If service robots are autonomously adaptive and can react with the environment in a way that is not predictable in detail, this raises the question of the creation of an independent legal "liability" of these novel mechanical "beings" which has to date only been discussed for software agents.

Ethical Perspective

From an ethical point of view, the focus is on the desirability of certain technical solutions regarding their *reasonability*. These questions will be discussed hereafter on the example of robots in caregiving/medical services.

Today, services in the field of caregiving, or medical care in general, are typically provided by human beings. However, the statistics for industrialized countries predict a demographic change which means that the number of people in need of care will be growing in the foreseeable future while the number of caregivers will decrease. Against this background it could be desirable for a society to develop service robots for care (Sparrow & Sparrow 2006). Their use can be planned to different extents, with the spectrum reaching from simple assistance in caregiving to "real" care robotics in the narrower sense.

Ethical questions on the desirability, which are connected to such scenarios, usually refer to the classical questions of ethics of technology. This is about the scientific reflection of moral statements which are often cited as arguments for the acceptance or the rejection of the use of technology. Cost-benefit considerations also play a role here. The questions are then answered with reference to procedural utilitarian, discursive, or participatory approaches. Such ethical considerations in the narrower sense form the standard repertoire of ELSI concepts which are also common for robotics and the used autonomous systems in parallel to on-going research (cf. e.g. Royal Academy 2009). A comprehensive ethical reflection also includes methodological questions aiming at the determination of what should be considered succeeding or even successful support, replacement, or surpassing of human performances, abilities, or skills. Then the design criteria for the description of robotic systems which replace human actors gain centre stage (cf. Gutmann 2010; Sturma 2003). The methodological reflection focuses on an equalization of human and machine. This is followed by the differentiation of human-machine, machine-human, machine-machine, and human-human interaction where a differentiation of interaction and interface could become relevant, terms that are often used synonymously (cf. Hubig 2008). Only such a clarification can provide information on the logical grammar of the "as-if" structure and thus the attribution of emotive, volitional, and cognitive terms to robotic systems. A systematic clarification of the logical structure of such equalizations is directly relevant for solving the above-mentioned ethical questions.

Questions of anthropological dimensions are directly associated, since services in the field of medicine/care are currently performed by humans, as stated above. Thus the introduction of technical systems replaces the human being in some areas, technical systems are increasingly involved in human actions, machines will act in the role of humans in an "as-if" mode.

This expansion of the ethical consideration which complies with the double meaning of ἔθος and ἦθος (Gethmann & Sander 1999, pp. 121ff.) finally allows asking for concepts of man which are – normally implicitly – invested in the construction of the respective technology.

This background is necessary to address issues which go beyond a purely syntactical understanding of technical systems and can be phrased in the following way, taking health care services as an example:

How can a successful care service be classified as "keeping the meaning?" Such a classification does not only require "technical specifications" but also a comprehensive description of the service provided – also considering, e.g., friendliness, care, etc.

How can this "successful service" be determined as being factually successful? Does this require a long dialogue between "receiver" and "provider" in the sense of a human-machine, machine-human or a parallel communication via human-human dialogues?

Psychological Perspective

The design of the "interface" between human being and robot is a central element of service robotics. The robot can be integrated as driver assistance system into the "total system automobile," a vacuum cleaning robot is a "faceless" rolling robot and there are also humanoid robot systems. Within these human-robot systems exists a clear assignment of functions of human and robot which answers the question which tasks are performed by the robot and which by the human being – from the psychological point of view one of the most important questions regarding the design. However, this division of tasks bears the risk that the human being is only taking over those (remaining) tasks which the robot cannot carry out. This question is also relevant in non-working contexts – i.e. in private life: Which tasks could and should remain with the human, which tasks should be taken over by the robot?

Depending on the general allocation of tasks between human and robot, (ergonomic) issues which can be assigned to the human-machine communication have to be dealt with from the psychological point of view. Concerning the dialogue properties of programmes, DIN ISO 9241 for example lists "suitability for the task," "self-descriptiveness," "controllability," "conformity with user expectations," "error tolerance," "suitability for individualization," and "suitability for learning." These issues also play an important role in service robotics, where decisions have to be taken that affect the handling and user-friendliness of the robot system.

When it comes to making a technical systems user friendly, the criterion of "intuitive" handling is of great relevance today, e.g. in the context of mobile phones. In the field of service robotics this issue gains a special relevance: The aspect of "intuitive" handling focuses on the "appearance" of the robot, which brings humanoid robot systems into play. People tend to personalize things and thus also technology. So the question is also how humanoid should a robot system be for a special task, which is, like in our example, a service task and being performed in peoples´ privacy. The hypothesis of "Uncanny Valley" (MacDorman & Ishiguro 2006) suggests that an appearance which supports cooperation can turn into an "eerie" perception, which is counterproductive for user friendliness.

The industrial psychological consideration suggested here puts special emphasis on the allocation of tasks between human and robot in the cognitive field. Basically, this is a question of sharing and interaction between human and artificial intelligence: When may and should the robot provide a service autonomously based on the assessment of a situation without having received specific instructions? When is it allowed to correct assumed mistakes in the action of humans without explicit order? This is a psychological issue since questions concerning the ability to judge and mental capability play a role here; however, it also touches the ethical and legal dimensions of technology assessment.

This multidisciplinary approach can still be extended. Socio-scientific aspects can be included (Böhle & Pfadenhauer 2011), for example with empirical studies, to systematically analyze the concrete acceptance on the part of those who provide the service and those who receive the service. This could especially be extended on the level of so-called sub-disciplines; their relevance for the subject is quite justifiable (Decker & Grunwald 2001).

Why Humanoid Robots?

Humanoid robots are robots whose shapes are modelled on the human body. They have two legs, two arms, a torso, and a head and also joints (legs, arms, shoulder) which are designed for movements similar to those of humans. Anthropomorphic robots are also of "human shape" in a literal sense. The publications on robotics do not make a distinction in the use of these descriptions. Literally translated, "android" robots or "gynoid" robots are of the shape of a man or a woman. In some publications these terms are used to describe humanoid robots whose appearance is "confusingly similar" to a man or a woman: "an artificial system defined with the ultimate goal of being indistinguishable from humans in its external appearance and behavior" (MacDorman & Ishiguro 2006, p. 298). Typically they are covered with a skin-like coating, are apparelled and make – often rudimentary – facial expressions. In contrast, humanoid/anthropomorphic robots can be identified as robots "at first glance":

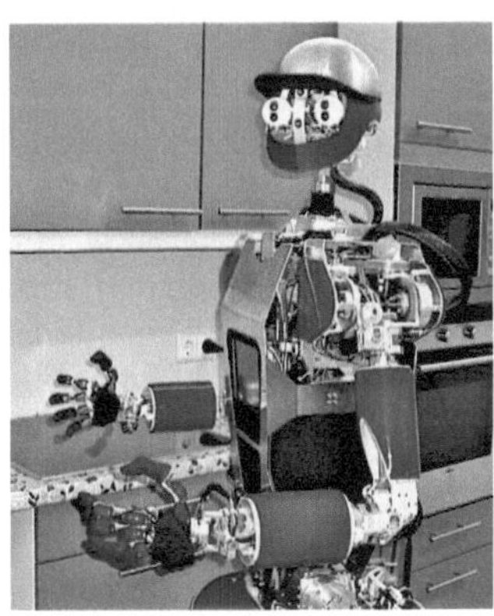

Figure 1: The humanoid robot ARMAR III (SFB 588) of KIT, Germany, by Jens Ottnad.

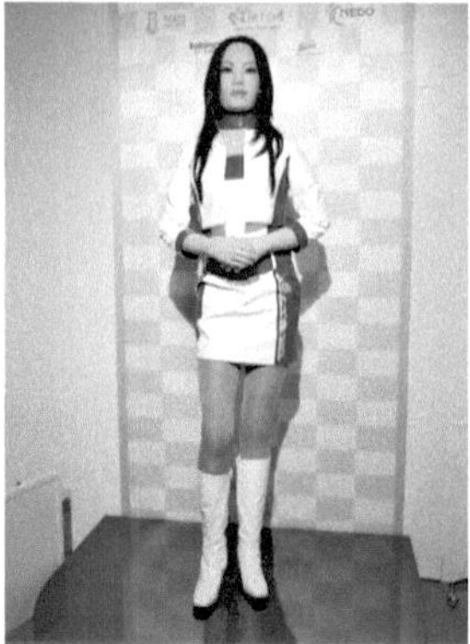

Figure 2: The "gynoid" robot ACTROID-DER, developed by KOKORO Inc. for customer service, appeared in the 2005 Expo Aichi, Japan. The robot responds to commands in Japanese, Chinese, Korean, and English.[4]

So the special humanoid shape has to be added to the general definition of "robot." The complexity which is requested in the definition of robots (see above) generally exists within humanoid robot systems, the descriptions in the humanoid robots catalogue on the web site of the SFB 588 "Humanoid Robots – Learning and Cooperating Multimodal Robots" may confirm this (SFB 588).

This robots catalogue can also be used as a first reference for the description of the status quo of humanoid robots research. First of all it shows that research institutes for the development of humanoid robots exist all over the world. They do research, also competitively, in similar fields; a closer look reveals the extremely modular structure of robotics research. They focus on the one hand on simple walking machines with two legs (e.g. Shadow Biped), on the other hand on robot heads with facial expressions and voice (e.g. Kismet). There are torso robots which are only human-like from the hip upwards and are either mounted on a rolling barrel or fixed on a table (e.g. ARMAR). Some robots are designed as high tech entertainers (flute player[5], dancer) or promote technology companies like Honda and Toyota. The basic paradigm of robotics in general is the fact that these modules can be integrated into a robotic overall system. Vice versa humanoid robots then represent that area of application of robotics where the "high end" products of different subfields of robotics research are in demand.

4 Photo by Gnsin, Wikimedia Commons, URL: http://de.wikipedia.org/w/index.php?title=Datei:Actroid-DER_01.jpg&filetimestamp=20110509163402 (accessed April 18, 2012)

5 See also the chapter "Understanding the Feasibility and Applicability of the Musician-Humanoid Interaction Research: A Study of the Impression of the Musical Interaction" (Jorge Solis & Atsuo Takanishi) in this book.

Language understanding, natural language, image recognition, learning, manipulation, etc. are the spearheads of the challenges in robotics research.

If we search for the reasons why robotic experts choose a humanoid shape for a robot system, we can find different answers which we will discuss in detail here.[6] Humanoid robots are built because the humanoid shape enables the robot to perform its task in a better way (1). This is another aspect in the ends-means context. Humanoid robots are built since it is an age-old dream of mankind to create artificial human beings (2); humanoid robots are built because we can thus learn something about human beings (3).

The Humanoid Shape as a Means to an End

This is based on the assumption that robots are employed as means to a specific end in some fields of application and that this end can be better achieved if the robot has a humanoid shape. The description of the Collaborative Research Center 588 (Humanoid Robots – Learning and Cooperating Multimodal Robots) considers the anthropomorphic shape as a prerequisite if a robot shares its activity space (e.g. kitchen) with a human being and they thus get in direct contact:

> "The aim of the project is to develop concepts, methods, and concrete mechatronical components for a humanoid robot which shares its activity space with a human partner. With the aid of this 'partially anthropomorphic robot system,' it will be possible to step out of the robot cage to realize a direct contact with humans."

The reason for the humanoid or "anthropomorphic" (as the SFB 588 calls it synonymously) shape is the fact that the robot has to act in an everyday environment that is designed for human beings. Sven Behnke phrases the goals for building humanoid robots in a similar way:

> "These efforts are motivated by the vision to create a new kind of tool: robots that work in close cooperation with humans in the same environment that we designed to suit our needs. […] Stairs, door handles, tools, and so on are designed to be used by humans. A robot with a human-like body can take advantage of this human-centred design. The new applications will require social interaction between humans and robots. If a robot is able to analyse and synthesize speech, eye movements, mimics, gestures, and body language, it will be capable of intuitive communications with humans. […] A human-like action repertoire also facilitates the programming of the robots by demonstration and the learning of new skills by imitation of humans, because there is a one-to-one mapping of human actions to robot actions. Last, but not least, humanoid robots are used as a tool to understand human intelligence." (Behnke 2008, p. 5)

Except for the last aspect (cf. the third reason: learning something about the human being), all criteria are aimed at the best possible completion of tasks – and thus to an ends-means relationship.

---

6 These reasons were already described in Decker 2010.

The first criterion refers to the environment in which human beings act. It is designed for human beings of average height. This includes doorways with a width of approx. 80 cm and a height of approx. 2 m, steps that can be climbed with human legs, i.e. are adjusted to the length of the leg and the knee joint, cupboards arranged in a way that they can be reached with a typical arm length, information signs at "eye level," most items have a handle which is optimized for the human hand, etc.

The second criterion refers to human intuition. The human being who acts in everyday life is not a robotics expert. However, he or she is used to "interact" with people. A humanoid shape takes advantage of this habitus. Even if one does not want to interact with a robot but only pass it in a safe distance, the shape of a humanoid robot indicates this distance. We keep to the custom of the safety distance between human beings ("a good length of an arm"). In a cooperation scenario, e.g. the joint carrying of an item, the human being normally takes over the "adaptive" part of the action. This adaptation is easier if the robot carries "like a human being." The general overall objective of a service robot is being operational in many fields and numerous action contexts.

> "We design humanoid robots to act autonomously and safely, without human control or supervision, in natural work environments and to interact with people. We do not design them as solutions for specific robotic needs (as with welding robots on assembly lines). Our goal is to build robots that function in many different real-world environments in essentially the same way." (Adams et al. 2000, p. 25)

Another question is how "human-like" should a humanoid robot be. For the aspects stated so far it is absolutely sufficient to have the shape of a human being, but look like a robot, i.e. "technically." A look at the gallery of the by now great number of humanoid robots reveals that they will hardly be mistaken for human beings. Geminoid, built by Hiroshi Ishiguro at the University of Osaka, is one example that also gained the attention of the media. Geminoid is the spitting image of Ishiguro. According to Ishiguro´s own statement he chose this appearance to be able to analyze the reactions of his students. So the question is: Should a humanoid robot be an exact copy of a human being (and thus "android/gynoid") to be able – and this is the difference to aspect 2, which focuses on "simulation" – to perform its tasks better? Concerning the first criterion, the fact that the environment is optimized for human beings, we could state that the human resemblance is not conducive. This is about similar shape and arrangement of joints. With intuitive handling, we have a different situation. We could say that the more humanoid a robot is the better works intuition. However, we also have to consider the question whether some part of the intuition might be "reversed" by a certain amount of irritation if the robot´s facial expressions "reveal" emotions (e.g. face robot Kismet and others). The reason for this irritation would be the fact that there is no equivalent to these human "feelings" in the robot´s control system (cf. above "Uncanny Valley"). But nevertheless, it is a necessary precondition for modern learning algorithms (see below) which pursue the aim that a robot learns "like a child" that humans treat them "like other human beings." This could in turn be supported by a "confusingly similar shape."

Creating Artificial Human Beings

The "dream of mankind" to create artificial human beings can only be briefly described here and without making a claim to be complete or describing all the details of the procedure. One possible starting point could be "Machine man" ("L'Homme Machine") by Julien Offray de La Mettrie (1709-1751); at least it is considered as visionary for AI and robotics research (e.g. Franchi & Güzeldere 2005, p. 39; Irrgang 2005, p. 28). De La Mettrie states: "So let us draw the bold conclusion that the human being is a machine..." (de La Mettrie 2001, p. 94) and

> "[The soul] is only a principle of movements or a sensitive material part of the brain which can be regarded as the main driving force of the whole machine, without risking being wrong."

These statements are a good example for the time of the mechanistic world view which was influenced by materialism when theories of "animal machines" were developed and also the human being was, according to de La Mettrie, described as self-governing machine.

This was accompanied by the discussion of the marriage of automata making. Vaucanson's flute player, tambourine player, and mechanical duck are regarded as masterpieces of automata making. In a brochure of an exhibition of that time the flute player was described as follows:

> "It is a man of normal height, who is dressed like a savage and plays eleven tones on the flute with the same lip and finger movements and the same breath as a living human being."

This description was complemented by the chronicler of the royal court, Duke of Luynes:

> "The air really comes through the mouth, and the fingers play. The fingers are made of wood with a piece of skin at the place where the holes are closed."

On the occasion of the 300th anniversary of Vaucanson's birth he was called the "father of mechanical creatures" in the category "On this Day" of the West German Broadcasting Corporation. There it said: "The audience is enthusiastic and people pay high entrance fees. The machine that looks like a human being seems to be alive." (WDR Stichtag 24.2.2009)

Even though Vaucanson wanted to achieve this appearance of being alive (to this end, the mechanical duck was extensively decorated with feathers), he also wanted to find out how man "functions." In the case of the flute player, it was about human breathing. Later he followed the plan to build an artificial human being as anatomic reproduction which should demonstrate how human viscera work (cf. the following third reason for building humanoid robots). But nevertheless, in general the perfect appearance of the living was in the focus of activities in that time. Baron von Kempelen widened this "more appearance than reality" by the aspect of thinking. His chess-playing automaton "The Turk" was hardly looking like a human being, but had a speech mechanism and was able to play chess – and was thus supposedly able to "think." The ability to play chess was realized by a Liliputian in the box of the automaton who moved the chess pieces with a mechanism.

The efforts of robotics researchers to build "android" or "gynoid" robots also fall in this category. In general, they are equipped with a silicone skin; so-called artificial muscles enable them to make different facial expressions. These facial expressions are based on human features and shall express different feelings. The robot face by Hara and Kobayashi (1995) is able to display or rather emulate six different moods. Since these moods are not connected to emotions inside the robot (like grief, joy, etc. which we attribute to human beings), it is again only a case of preserving and optimising a preferably human appearance. But nevertheless, the expression of these moods also pursues certain ends which, vice versa, improve the functionality of the robot (cf. first reason: ends-means relationship).

The realization of the age-old dream of mankind to create artificial human beings is explained in current publications on robotics. The quotation "The humanoid robot has always been our dream" (Yokoi et al. 2004) may be just one example here. The authors use it as first sentence in their introduction, and then they dedicate themselves completely to technical questions.

Learning from Human Beings and for the Understanding of Human Beings

In general, bionics pursues the goal of learning from nature. There are different definitions of bionics, some of them also bring normative aspects into play – for example that bionics is also said to include the promise to allow for solutions which are more ecologically acceptable and societally unproblematic (Grunwald 2009, pp. 19ff). Concerning the core of bionics, most definitions agree that findings from the study of a biotic model influence the development of a technical artefact and thus contribute to the technical problem solving (Oertel & Grunwald 2006, pp. 24ff.). A bionic approach seems to be especially suitable for humanoid robotics. To achieve a humanoid gait it can make sense to be guided by the "original." But also for "normal" robotics and the robot "arms" developed, the relation between own weight and "payload" of the human arm is unmatched. Therefore it is not surprising that the objective of the BIOROB joint research project is building a robot arm based on the human model (see Biorob). One "side effect" – which might become the main objective in connection with service robotics – is the fact that the arm is flexible and therefore carries much less risks of injury for the people in contact with the robot. This makes the man-robot interaction safer (Klug et al. 2008; Fraunhofer 2010). The problem of these "promises" of bionics and the inherent relation between technology and nature will not be discussed in detail here. For this argument it has to be noted that a bionic approach of technology design can be especially reasonable for humanoid robots.

Vice versa this "technified glance" (Grunwald 2009) which bionics throws at the natural human being brings about new knowledge on man. Rolf Pfeifer puts this end into the focus of his research: With the help of robot systems he develops hypotheses for biological research. Artificial ants may serve as an example here; with their help Pfeifer developed the hypothesis that natural ants are able to solve simple arithmetic problems in their neurons. Desert ants orient themselves on the sun´s polarization pattern. They possess three "sensors" with which they observe the polarization pattern in different directions. With the help of a robotic ant which was equipped with an identical ("technical") setting, Pfeifer and his team were able to demonstrate that the mere recognition of the polarization pattern did not lead to a real ant´s precision in determining the direction. Only if the three pieces of "information" are linked (the absolute difference of two sensor signals are deducted from the third one), the robot ant is able to achieve the same directional precision as the "original." For Pfeifer, the result of this experiment is the hypothesis that desert ants must have a possibility (e.g. via neurons) to make such calculations. This hypothesis would have to be verified by biologists (interview with Rolf Pfeifer (KI 2003)). According to his statement in Swiss TV (SF1 in 2008), Pfeifer builds "robots to understand how human beings function." Adams et al. (2000, p. 25) argue in a similar way in reference to human intelligence:

> "Robotics offers a unique tool for testing models drawn from developmental psychology and cognitive science. We hope not only to create robots inspired by biological capabilities, but also to help shape and refine our understanding of those capabilities. By applying a theory to a real system, we test the hypotheses and can more easily judge them on their content and coverage."

If these three reasons for building humanoid robots are considered together, it soon becomes apparent that these "motivations" also show certain overlaps and that robotics researchers (like Adams et al. 2000, p. 25) each name a combination of two or even all three reasons. So simulating a "human" machine with the focus on the externally visible performance – which was observed in the tradition of automata making – becomes relevant again if the similarity to human beings shall serve to achieve a certain aim. At least this improvement in achieving the aim justifies the humanoid design. The work of Cynthia L. Breazeal may serve as an example here, where the facial expressions of a robot are used to keep the person who is speaking to the robot at a distance that is ideal for the robot. The robot regulates the interaction:

> "Regulating interaction via social amplification. People too distant to be seen clearly are called closer; if they come too close, the robot displays discomfort and withdraws. The withdrawal moves the robot back only a little physically, but is more effective in signalling to the human to back off. Toys or people that move too rapidly cause irritation." (Breazeal 2000, p. 238)

Something similar can be stated for reasons one and three. If hypotheses how human cognitive performances could possibly be achieved are developed in the field of humanoid robotics this is just one side of the coin. In a second step these "technically verified hypotheses" will be used as technical approaches for the construction of robots. Adams (et al. 2000, p. 25f.) combine just these two motivations in the principles behind their methodological approach to develop autonomous robots which are able to act in an environment that is adapted to human beings and are also able to socially interact with human beings.

Finally I will also point out the overlaps between reason one and three. According to the sources, humanoid automata were mainly intended for entertainment; therefore the automata makers were travelling from town to town to display their automata. However, it was also emphasized that the understanding of the viscera of humans and animals was also an issue. Vaucanson´s human and animal automata are examples for this.

Turning now to a robot´s ability to learn, this can be justified with all three motivations. On the one hand, the simulation of a humanoid robot is easily "seen through" if the robot is not able to acquire new skills by learning. This applies both for communication in natural language and for the ability of manipulation, i.e. if we want to show the robot how something works. On the other hand, cognitive sciences are also interested in an experimental support of the modelling:

> "Different methodological approaches exist in cognitive sciences, especially the experimental study of human cognition in psychology and of its neuronal substrates in neuroscience, it´s modelling in formal description and computer simulation in computer sciences/mathematics and the formal analysis of the results of cognitive processes in linguistics." (Interdisziplinäres Zentrum für Kognitive Studien [Translated by the author, Michael Decker.])

The hypotheses on learning abilities by Pfeifer and Adams et al. mentioned above complement the described simulations of computer sciences.

And finally, as third aspect, the learning ability is a central requirement for service robots because in general we have to assume that they have to be able to explore their environment on their own. It is quite simply not possible to program a robot top-down for numerous tasks in an unknown environment. This points to the problem of scaling up:

> "How can we progress from the 20-40 behaviour patterns common today to thousands, millions or even more? This is the scaling-up problem which to date has not been solved by a convincing concept." (Christaller et al. 2001, p. 73)

Vice versa, the robot´s ability to learn is also a decisive requirement for the robot users who are not experts in robotics. It is hardly conceivable that a robot user is programming a robot at home on his own, even with a detailed manual.

Conclusions

A technology assessment of service robotics can take up the results of TA studies on robotics in general (Kündig & Bütschi 2008; Royal Academy 2009; Lau et al. 2009). Concerning learning robot systems, Christaller et al. (2001) suggested to distinguish them from other robot systems since there might be a grey area between product liability and owner liability. Mathias (2004) also pointed out this "responsibility gap." As we have to assume a high flexibility of service robots in private use, this issue is of central importance. With regard to the hierarchy of control in the cooperation between robots and human beings, we can also revert to existing recommendations:

> "In the contexts of robotics we have to adhere in principal to the competence of people to define the aims. The related interdiction of instrumentalization has to be taken into account when the relevant hierarchy of decisions is established.
>
> The configuration of the human-machine interface and/or program control is of high relevance for the technical realization of the decision-making competence. People can only take over the responsibility for the proper functioning of robots if the robots can be controlled in terms of transparency, predictability, and influence.
>
> In all cases where robots get leeway for decisions, it is recommended to inform the people involved and ask for their explicit or implicit consent. Especially in the case of medical treatments and care the refusal of this consent shall be equivalent to a veto." (Christaller et al. 2001, p. 220)

This recommendation is relevant for service robotics since services involving human beings can have a special quality of "cooperation," e.g. when it comes to taking medicine where the robot might admonish the patient. Such an admonishment from a human caregiver is nothing unusual and clearly within the scope of his or her duties. Not complying with these instructions can lead to sanctions, if the patient is not willing to cooperate. Would such instructions from a service robot also be accepted?

This recommendation can also be taken as an example for the interdisciplinary intertwining of disciplinarily established arguments. Using the ethical argument of interdiction of instrumentalization, the requirements for the technical implementation of the human-machine interface are formulated and additionally supported by the legally justified argument that the responsibility for actions still has to be attributable to the initiator. The consequence which is, especially in the care sector, related to this "right of veto" then also implies economic consequences since this right of veto cannot be realized without additional costs and therefore has to be included in a cost-benefit calculation.

For a TA of service robotics, the shape of robot systems has to be moved into the centre of attention like it has never been the case before. While robot systems have to date been mainly designed with regard to functional criteria and aesthetic aspects played a minor role (again expansion robots may be used as plausibility argument here: the look of a deep sea robot or a robot for the inspection of nuclear power stations is subordinate), the appearance of service robots will be of increasing importance. Here we will have to distinguish especially between purely creative aspects concerning the appearance of robots which do not influence its functions at all, and those improving the functionality which also have to be assessed on the basis of ends-means aspects. In this context we will have to examine especially humanoid, "android," and "gynoid" robot systems where we can find on the one hand concepts that support the intuitive use of robot systems. In doing so, we have to weigh up the possible advantages against an eerie appearance ("Uncanny Valley") and against suspected manipulation, e.g. expressed by Christaller et al. (2001, p. 218).

For some time now service robots have been predicted to have an economic potential similar to industrial robots which have become an integral part of the automotive industry, just like metalworking, plastics, rubber, timber, and furniture industry. From the perspective of research of the impacts of technology, this includes questions that go beyond the mere ends-means relationship. Both in public and in private, the robots may encounter "third parties" which do not expect a service robot "in action" and therefore have to be taken into consideration as unpredictable source of danger. In addition, it has to be analysed in every context of actions which aspects of the service provided by a human being can be/could have been conferred on the service robot while "keeping the meaning." Due to the strong relevance of the contexts, a TA oriented on case studies seems advisable, which was suggested by Decker et al. (2011) and has now been started as a project at the Europäische Akademie GmbH.

References

Adams, B., C. Breazeal, R. Brooks & B. Scassellati 2000: "Humanoid Robots: A new kind of tool." In: IEEE Intelligent Systems 15(4), pp. 25-31.

Behnke, S. 2008: "Humanoid Robots – From Fiction to reality?" In: Künstliche Intelligenz (4), pp. 5-9.

Bleicher, A. 2010: "The first Gulf Spill's lessons for Robotics. The demands on the largest underwater robotics armada ever fielded show that ROV's need better automation." In: IEEE Spectrum August 2010, pp. 9-11.

Bleicher, A. 2011: "Gulf Spill One Year Later: Lessons for Robotics" (Available at: http://spectrum.ieee.org/robotics/industrial-robots/gulf-spill-one-year-later-lessons-for-robotics [accessed August 25, 2011])

Biorob: http://www.biorob.de/ (accessed August 24, 2011)

Böhle, K. & M. Pfadenhauer 2011: Schwerpunkt: Parasoziale Beziehungen mit pseudo-intelligenten Softwareagenten und Robotern. Technikfolgenabschätzung – Theorie und Praxis (1), 20. Jg.

Breazeal, C. 2000: Sociable Machines: Expressive Social Exchange. PhD-Thesis. Massachusetts Institute of Technology.

Christaller, T., M. Decker, J.-M. Gilsbach, G. Hirzinger, K. Lauterbach, E. Schweighofer, G. Schweitzer, D. Sturma 2001: Robotik. Perspektiven für menschliches Handeln in der zukünftigen Gesellschaft, Berlin & Heidelberg.

Decker, M. 2000: "Replacing Human Beings by Robots. How to Tackle that Perspective by Technology Assessment?" In: Grin, J. & A. Grunwald (eds.) 2000: Vision Assessment: Shaping Technology in 21st Century Society. Towards a Repertoire for Technology Assessment, Heidelberg et al., pp. 149-166.

Decker, M. & A. Grunwald 2001: "Rational Technology Assessment as Interdisciplinary Research." In: Decker, M. (ed.) 2001; Interdisciplinarity in Technology Assessment. Implementation and its Chances and Limits, Berlin, pp. 33-60.

Decker, M. 2010: "Ein Abbild des Menschen: Humanoide Roboter." In: Bölker, M., M. Gutmann & W. Hesse (eds.) 2010: Information und Menschenbild, Berlin & Heidelberg, pp. 41-62.

Decker, M., R. Dillmann, M. Fischer, M. Gutmann, I. Ott & I. Spiecker gen. Döhmann 2011: "Service-Robotics: do you know your new companion? Framing an interdisciplinary technology assessment." In: Poiesis & Praxis. International Journal of Ethics of Science and Technology Assessment 8, pp. 25-44.

De la Mettrie, J. O. 2001: Der Mensch eine Maschine, Stuttgart.

Franchi, S. & G. Güzeldere (eds.) 2005: Mechanical Bodies, Computational minds. Artificial intelligence from automata to cyborgs, Cambridge (MA).

Fraunhofer 2010: Deutscher Zukunftspreis 2010 für Festo und Fraunhofer IPA: Raffinierter Rüssel: http://www.fraunhofer.de/presse/presseinformationen/2010/12/Deutscher-Zukunftspreis-2010.jsp (accessed April 18, 2012)

Gethmann, C. F. & T. Sander 1999: "Rechtfertigungsdiskurse." In: Grunwald, A. & S. Saupe (eds.): Ethik in der Technikgestaltung. Praktische Relevanz und Legitimation, Berlin & Heidelberg.

Grunwald, A. 2009: Bionik: "Naturnahe Technik oder technisierte Natur?" In: Müller, M. C. (ed.) 2009: Der Mensch als Vorbild, Partner und Patient von Robotern. Bionik an der Schnittstelle Mensch-Maschine, Loccum.

Gutmann, M. 2010: "Autonome Systeme und der Mensch: Zum Problem der medialen Selbstkonstitution." In: Selke, S. & U. Dittler (eds.): Postmediale Wirklichkeiten aus interdisziplinärer Perspektive, Hannover, pp. 130-148.

Hara, F. & H. Kobayashi 1995: Use of Face Robot for Human-Computer Communication. Systems, Man and Cybernetics. Intelligent Systems for the 21st Century. IEEE International Conference.

Hubig, C. 2008: "Mensch-Maschine-Interaktion in hybriden Systemen." In: Hubig, C. & P. Koslowski (eds.) 2008: Maschinen, die unsere Brüder werden, München, pp. 9-17.

Hüttenrauch, H. 2006: From HCI to HRI: Designing Interaction for a Service Robot, Stockholm.

IFR: "Service Robots" (http://www.ifr.org/service-robots/ [accessed Mai 20, 2010])

Interdisziplinäres Zentrum für Kognitive Studien: "What is cognitive science?": http://www.kogni.uni-potsdam.de/forschung/kognitionswissenschaft.html (accessed April 18, 2012)

Irrgang, B. 2005: Posthumanes Menschsein? Künstliche Intelligenz, Cyberspace, Roboter, Cyborgs und Designer-Menschen. Anthropologie des künstlichen Menschen im 21. Jahrhundert, Stuttgart.

Jakuba, M., J. Kinsey, D. Yoerger, R. Camilli, C. Murphy, D. Steinberg & A. Bender 2010: "Exploration of the Gulf of Mexico Oil Spill with the Sentry Autonomous Underwater Vehicle." (Available at: http://www.isr.uc.pt/WREM2010/Files/Papers/JakubaWREM2010.pdf [accessed August 25, 2011])

KI 2003: "Interview: Prof. Dr. Rolf Pfeifer über die 'Neue Künstliche Intelligenz.'" In: Künstliche Intelligenz 1/03, pp. 50-54.

Klug, S., T. Lens, O. von Stryk, B. Möhl, A. Karguth 2008: "Biologically Inspired Robot Manipulator for New Applications in Automation Engineering." (Preprint of the paper which appeared in the Proc. of Robotik, Munich, Germany, June 11-12, 2008)

Kündig, A & D. Bütschi 2008: Die Verselbständigung des Computers, Zürich.

Lau, Y. Y., C. van´t Hof & R. van Est 2009: Beyond the Surface – An Exploration in Healthcare Robotics in Japan, The Hague, Rathenau Instituut TA report.

MacDorman, K. F. & H. Ishiguro 2006: "The uncanny advantage of using androids in cognitive and social science research." In: Interaction Studies 7:3, pp. 297-337.

Matthias, A. 2004: "The responsibility gap: Ascribing responsibility for the actions of learning automata." In: Ethics and Information Technology 6, pp. 175-183.

Oertel, D. & A. Grunwald 2006: Potenziale und Anwendungsperspektiven der Bionik. Arbeitsbericht Nr. 108, Büro für Technikfolgen-Abschätzung beim Deutschen Bundestag, Berlin.

Royal Academy 2009: Autonomous Systems: Social, Legal and Ethical Issues. The Royal Academy of Engineering, London. (Available at: http://www.raeng.org.uk/societygov/engineeringethics/pdf/Autonomous_Systems_Report_09.pdf [accessed April 18, 2012])

Schraft, R. D., M. Hägele, & K. Wegener 2004: Service Roboter Visionen, München.

SFB 588: http://www.sfb588.uni-karlsruhe.de/old/textdateien/ziele_frame.html (accessed August 24, 2011)

Sparrow, R. & L. Sparrow 2006: "In the hands of machines? The future of aged care." In: Mind Mach 16, pp. 141-161.

Spiegelonline 04.06.2010: http://www.spiegel.de/wissenschaft/natur/0,1518,698667,00.html (accessed June 6, 2011)

Stöcker, C. 2011: "Reaktorkatastrophe in Fukushima. Roboter für den Höllenjob." In: Spiegelonline, 19.3.2011: http://www.spiegel.de/netzwelt/web/0,1518,751880,00.html (accessed June 6, 2012)

Sturma, D. 2003: "Autonomie. Über Personen, künstliche Intelligenz und Robotik." In: Christaller, T. & J. Wehner (eds.): Autonome Maschinen, Wiesbaden, pp. 38-55.

WDR Stichtag 24.2.2009: http://www1.wdr.de/themen/archiv/stichtag/stichtag4170.html (accessed August 25, 2011)

World Robotics 2009: "IFR Statistical Department, hosted by the VDMA Robotics + Automation association." (http://www.worldrobotics.org [accessed August 25, 2011])

Yokoi, K, F. Kanehiro, K. Kaneko, S. Kajita, K. Fujiwara & H. Hirukawa 2004: "Experimental Study of Humanoid Robot HRP-1S." In: The International Journal of Robotics Research 23(4-5), pp. 351-362.

Popular Culture and Music Robots

Robots in Japanese Popular Culture

Maika Nakao

Introduction: Robotics and Popular Culture

Japan made significant progress in robotics during the latter half of the 20th-century. In the 1960s, Ichiro Kato of Waseda University launched the first robotics group in the nation. Since that time, there have been significant developments in robotics in Japan, especially in humanoid robotics, both in academic and industrial circles. For example, the Advanced Step in Innovative Mobility (ASIMO) robot, a bipedal robot created by Honda, was first announced in 2005, and in 2006, Hiroshi Ishiguro at Osaka University announced a humanoid robot called "Geminoid," whose appearance resembles Ishiguro himself. Both these robots surprised people with their advanced performance. In addition to these examples, there are many other groups engaged in robotics in the academic and industrial circles of Japan.[1]

This raises the question as to why Japan has made such significant advancement in the development of robots. In order to understand robotics in Japan, we cannot overlook the nation's cultural background.[2] The primary feature of Japanese robotics is that its development has been strongly influenced by the country's popular culture, wherein robots are a representative character. For example, ASIMO's creator Masato Hirose said that he created the robot by imagining Astro Boy, one of the most famous Japanese anime characters and a robot. Hirose was ordered by Honda to create an Astro Boy, and he created ASIMO.[3] Robotics researchers have not only been inspired by anime but have also appealed to anime culture. For example, The National Institute of Advanced Industrial Science and Technology (AIST) created the robot "HRP-4C," which resembles a cute female (nymph robot).[4] Dr. Kajita and Hirukawa at AIST stated that they used a female face in order to appeal to the media.[5] Accordingly, robotics in Japan cannot be adequately explained without referring to popular culture as

---

1 There is a comprehensive guidebook on Japanese robotics including popular culture: Hornyak 2006.

2 There are several studies that seek the background of the different attitudes that Japan and Western countries have towards robots, for example MacDorman et. al. 2009. Religious and cultural reasons have often been used to describe these different attitudes.

3 Hirose talked about this story in various media, for example Hirose 2002.

4 Press Release of AIST, March 16, 2009: http://www.aist.go.jp/aist_j/press_release/pr2009/pr20090316/pr20090316.html (accessed April 15, 2012)

5 ROBOT WATCH article of July 16, 2009, reported by Kazumichi Moriyama http://robot.watch.impress.co.jp/docs/column/review/20090716_299492.html (accessed April 15, 2012)

well. Steffi Richter, a German scholar who teaches Japanese studies in Leipzig University, claims that more than in any other country, in Japan, science fiction, including monsters and robots, penetrates both daily life and the public sphere (including academia) (Richter 2010, p. 175).

Japanese popular culture has been garnering considerable attention from around the world and has also been the topic of recent scholarship on Japanology. One of the main questions is what makes Japanese popular culture so distinctive. Scholars of Japanese culture, such as Susan J. Napier, have indicated that historical events, such as being defeated in WWII and the nation´s subsequent rapid economic growth, have had a significant impact on Japanese popular culture. In other words, Japanese history has made Japanese culture unique (Napier 2005).

Based on this stated view, this paper will examine the representations of robots within a historical context, which will demonstrate the complex relationship between human and technology, self and other, and subject and object. The robots in popular culture exemplify the problem of Japanese identity being closely connected to technology. This paper reveals the problem that underlies Japanese robotics and popular culture. The first section discusses historical discourse on robots and indicates the problem of Japanese identity. The next section focuses on several representative robots in popular culture and explores how the problem of identity is mirrored or reflected in robots.

Discourses on Robots (Robots and Japanese Traditional Culture)

The most famous manga in postwar Japan and representative of robots is Osamu Tezuka´s Astro Boy, which was first published in 1952 and subsequently broadcast in 1963. Since then, Astro Boy has been considered the origin of Japanese robotics. Astro Boy has been used as an image by several companies and institutions. It seems that Astro Boy embodies the Japanese dream of technology. Astro Boy´s fictional birthday is April 2003; around this time, various ceremonies were held, and a new Astro Boy animation series was released. A website created by one of the largest newspaper companies in Japan displayed a countdown to his birthday.[6] The Takadanobaba Station, which is the location of the animation studio that produced Astro Boy, adopted as its departure announcement music the theme song from Astro Boy. People have long compared Astro Boy to contemporary robots, speculating as to what modern robotics can achieve (Ito 2010c, p. 63).

There are several explanations for the popularity of Astro Boy and other robots in Japan. For example, the editor´s note of a roundtable "Where do we go from here?" in *The Robot Chronicles* (2003) argues that, "The Japanese are fascinated with robots as if they were seeking therapy in the AIBO and it is a fact

---

6 For further discussion of Astroboy´s birthday, see Ito 2010a and 2010b.

that this national characteristic became a driving force in creating the robot powerhouse of Japan."[7] Cultural anthropologist Akio Yamaguchi made the following claim:

> "What is interesting to me is that Western people are not familiar with robots. There are many supermen who are close to God, but very few characters such as Astro Boy. I think there is a resistance to depicting robots as central characters (in the West). Robots were seen as just a tool and not seen as coexisting with humans. But Osamu Tezuka created a manga hero, which started a character of robot. The Japanese have their own tradition that drew on dolls, such as *Bunraku* in which controller and doll are united. (abbr.). Tezuka used a Japanese sensibility in which robots and dolls can easily live together with humans and he created a hero such as the Astro Boy." (Yamaguchi 1990, pp. 237-238)

Yamaguchi emphasized that the Japanese tradition is well-suited robots. Ichiro Kato, who first started robotics research in Japan in the 1960s, indicated that *Karakuri* and robots share a feature, that is, they are both deformed. *Karakuri* is the Japanese traditional mechanical doll. People have believed that robots could be linked to *Karakuri* because it represented Japanese technical excellence. In this way, Japanese people have tended to reference traditional culture to explain their fascination with robots.

Not only have the Japanese emphasized the cultural background of robots, they have also emphasized the cultural background of Astro Boy. Tezuka, the creator of the Astro Boy, is considered the father of Japanese manga culture. It was previously argued that Japanese Manga culture was imported from US comics after WWII and that Pinocchio and Mickey Mouse were models for Astro Boy. However, more recently, the Japanese wartime culture has been examined as having influenced Tezuka's Astro Boy rather than the West.[8]

Historian Kenji Ito criticized the tendency to invoke Astro Boy as a reminder of Japanese traditional culture.[9] However, postwar Japanese people used a historical explanation, such as *Bunraku* and *Karakuri*, to illustrate the robot culture. At this point, the question arises of why historical roots were used to ex-

7 The Robot Chronicles: Catalog of an exhibition "Tetsuwan Atom no Kiseki ten" (Asahi shinbunsha 2002), p. 107.

8 For example, the book "Tanjo! Tezuka Osamu" (Shimotsuki 1997) attempted to look at the Japanese cultural background of Astro Boy. On the other hand, manga critiques problematize Tezuka's influence because later anime writers could not get beyond Tazuka and the said "Tezuka's circle" (Ito 2005).

9 Kenji Ito studied popular images of robots in the 1930s and 1950s in Japan and showed that although the robot in prewar Japan was conceived as having a tool-like existence, which could be utilized against the Japanese, in postwar Japan, the robot was conceived as a friend to humanity and a guardian of peace and democratic values. Then Ito claimed that these different images of the robot reflect society's perceptions of science and technology during these eras. Therefore, it can be said that the experience of WWII changed the popular images of robots or science and technology (Ito 2003).

plain robot culture. In order to think about this situation, we need to consider the Japanese problem of identity that is closely connected to science and technology. The discourse of Astro Boy reveals the complex relationship between technology and national identity.

Science and Technology, and Japanese Identity

The tendency to connect robot origins with Japanese traditional culture cannot be illustrated without explaining the problem of Western science and technology being introduced in Japan; the problem of science and technology and national identity arose with the Meiji restoration.[10] From the Meiji Restoration, Japan stepped toward the foundation of science and technology by importing ideas from Western countries. At the same time, Japanese intellectuals and politicians regarded science as incompatible with the Japanese spirit. The Japanese people of the Meiji era used the term "wakon yosai," which means Japanese spirit and Western learning. Science and technology were considered useful pursuits that could be reconciled with Japanese traditional culture. However, the Japanese needed confidence in their ability to overcome the West, and intellectuals needed to understand national identity in a manner compatible with science (Calichman 2008). During the 1930s, when Japan sought to expand its reach and launched wars of imperialism, there were several studies that attempted to show the greatness of the Japanese people and sought to express their national identity; for example, Hajime Tanabe discussed the need to synthesize the Japanese spirit and scientific spirit in 1936 (Tanabe 1936, p. 18). Some intellectuals, such as Hiroto Saigusa and Yoshio Mikami, revealed Japan´s own history of science (for example, Japanese mathematics Wasan) and attempted to establish a national identity.[11] Kazuo Fukumoto, a famous Japanese Marxist thinker, studied the history of *Karakuri* in 1944 (Fukumoto 2008). In this sense, we can grasp the discourse of Japanese technology, such as *Karakuri,* within the context of the struggle for a Japanese identity.

As opposed to *Karakuri*, the robot was a Western notion; historically, the notion of the "robot" gained global currency in the 1920s. It was in 1923 that Čapek´s play, "RUR Rossum´s Universal Robots," which first coined the term "robot," was introduced to Japan. After that, the term "robot" was used frequently in Japan; the robot was connected with the image of a mechanized civilization as represented in Fritz Lang´s "Metropolis" (Yoshida 2003). The popular image of the robot represents people´s anxiety toward such a civilization.[12] The term "robot" was also used for humans who had lost their political subjectivity. In 1931, when the first robot boom occurred in Japan, the politician and future

10 In Japan, "science" and "technology" were translated as "kagaku" and "gijutsu," and they are often described as overall "kagakugijutu" (Suzuki 2010).

11 For further discussion of this issue, see Mizuno 2009.

12 As for images of robots in prewar Japan, see Inoe 1993 and Yoshida 2003.

Prime Minister Ichiro Hatoyama criticized the commission cabinet as a "robot cabinet" and made the following statement: "the Hamaguchi cabinet after Osachi Hamaguchi is a headless cabinet. Nowadays 'robot' is attentive to what is being said ('robot' became a popular word), this is a 'robot' cabinet." (Inoue 2007, p. 101). Thus, the image of the robot was not a positive one and was sometimes even pejorative in prewar Japan.[13]

However, robots depicted a familiar existence comparable to humans in postwar Japanese popular culture. The next section observes several representative works on robots and examines the indissoluble connection between robots and humans.

Robots in Popular Culture

The most famous manga in postwar Japan and a representative work of robots is Osamu Tezuka's Astro Boy, which was first published as a manga in 1952 and then broadcast in 1963. In Tezuka's story, Astro Boy was created by Dr. Temma in the image of his son who was killed by a traffic accident. Although he is robot, his appearance is close to that of a human child, and he speaks and thinks like a human child. In fact, Tezuka wrote another manga, Ambassador Atom, on which Astro Boy is based. In Ambassador Atom, the original character of Astro Boy is nonhuman with an expressionless face and no emotions. Then Tezuka's editor advised him to change the nature of the character, and Astro Boy, who has weaknesses and feelings like humans, was born. Preserving the features of the original work, Astro Boy is an "ambassador of peace" and fought for humans. Astro Boy work as a gatekeeper of peace and democracy, which could be described as components of postwar ideology. In addition, Astro Boy is an ambassador, who will connect the US and Japan through "peaceful" nuclear technology. Astro Boy is driven by nuclear power, which can be seen as reflecting the positive reception of nuclear power in postwar Japan, which recalls Eisenhower's 1953 "Atoms for Peace" campaign. Although Astro Boy is apparently seen aligned with the postwar ideology of peace and democracy, he is distressed about several issues in the story.[14] The story depicts several ethical themes that can be discussed as the robo-ethics. For example, he is distressed and becomes sad because he does not have a "mind." However, can it really be said that he,

---

13 According to Haruki Inoue, who studied the acceptance of robots in pre-war Japan, there was a robotization of human beings in wartime Japan: "During the wartime, there seems to be a robotic disposition to human existence." This quote suggests that humans did not think independently and got carried away with the mood of the war as though they were robots (Inoue 2007, p. 433).

14 Although the story is not just a positive view of science and technology, Tezuka later (1980s) commented that Astro Boy did not reflect a positive view of science and technology; the appearance of Astro Boy can be seen as the symbol of positive acceptance of nuclear energy in postwar Japan (Tezuka 1997, p. 73).

who is distressed and has feelings, really does not have a mind? Thus, Tezuka described the delicate problem in Astro Boy´s story.

There is another robot anime entitled "Tetsujin 28-go," known as "Gigantor" in some countries, which was serialized starting in 1956. The story is set in Japan between 1955 and 1964, and follows a young boy named "Shotaro Kaneda," who controls a giant robot named "Tetsujin 28-go." Although the story is set in postwar society, it is affected by the past war; Shotaro´s late father originally built the robot as a special weapon to be used during the Pacific War. Then the robot appeared in postwar society; by controlling this giant robot, Shotaro prevails over criminals and promotes peace. The Tetsujin 28-go, which was created during the war, keeps postwar society safe. This setting can be seen as the appearance of the historical introduction into understanding the roots of the robot, as previously mentioned. The author Yokoyama experienced the War as a child, and he reflected on his experience in his work.

Thus far, we have seen the famous first-generation of robot anime that appeared during the 1950s and 1960s. Astro Boy and Tetsujin 28-go are the representative robots of the first-generation of post-war anime, with each being a distinct type of robot: the autonomous robot and the controlled robot. In this anime, robots are not against humans. In the 1970s, a genre of robot anime appears in which humans get inside and embody large robots, which was called "powered suit robot," which can be described as a fusion of the autonomous and the controlled robots, Astro Boy and Tetsujin 28-go.

The first type of such robot anime, written by Go Nagai, was Maginger Z, known as "Tranzor Z" in some countries. The manga version was serialized in the weekly magazine Shonen starting in 1972, and the TV anime appeared in 1973. In this story, a 16-year-old high school student, named "Koji Kabuto," climbs into a gigantic super robot, Mazinger Z, which has been created by Professor Juzo Kabuto in order to fight against the large robots created and controlled by the evil scientist Dr. Hell, who aims for world conquest. To prevent Dr. Hell from conquering the world, Dr. Kabuto´s grandchild Koji puts Mazinger Z on his head and controls it. Koji´s family name "Kabuto" means "helmet" in Japanese. The author Go Nagai, reveals that he had always loved Astro Boy and Tetsujin-28 go as a child, and he wanted to make his own robot anime. He got the inspiration for a giant robot that could be controlled from the inside like a car while stuck in a traffic jam (Hornyak 2006, p. 60). This idea can be seen as reverse of Marx´s notion of "alienation." In contrast to being alienated from technology, Man embeds himself in technology, which gives technology a human face. Mazinger Z is constructed with a fictitious metal called "Super-Alloy Z," which is forged from Japanium, a new element found only in the sediment of Mt. Fuji and driven by photon force energy. Mt. Fuji is the highest mountain in Japan and sacred icon for the Japanese.

The TV anime series Mobile Suit Gundam, which was first published as a TV anime in 1979, is regarded as the standard of Japanese anime. The story is set in a future world, a half-century after Man has begun immigrating cosmo-space; each cosmo-space is said to be a space colony. The most distant colony, the Principality of Zeon, has attempted to become independent from Earth. Then a child named "Amuro Ray" is accidently selected as the pilot for the robot Gundam. The story describes children′s difficulty and how such problems are overcome through their experience of meeting, fighting, and bidding farewell. A recurring theme is that the concept of justice is not clear between two opponents, which makes the story complicated and deep.[15] Mobile Suit Gundam also shows the empowerment of children by robots. Technology and Man are gradually harmonized, and the distinction between technology and humanity, between right and wrong becomes unclear.

In 1995, another significant robot anime appeared: The Neon Genesis EVANGELION was broadcast on TV from 1995 to 1996. The story is set in 2015, after a catastrophe called the "Second Impact," and the earth′s population has been decreased by half. Mankind is subject to attack by unknown enemies called "angels." In order to resist the angels, humans created robots called "EVANGELIONS." Specially selected boys and girls, who are fourteen years old, can pilot the EVANGELIONS; they are put inside the robots and then control them. The EVANGELION is not simply a machine but an actual, living entity. The pilots are selected from motherless children in order to increase the synchronization level with the EVANGELION, into which the pilot′s mother′s soul is loaded. The level of synchronization between the pilots and EVANGELION is one of the central matters of the story. When pilots are mentally unbalanced, the level of synchronization is reduced. In a sense, this story can be said to focus on the inner psychological world of children: their fragile and innocent psychology. Thus, we can see EVANGELION not simply as a machine but a living "creature."

In all of these anime, the central characters are children, who have an innocent heart and are weak in their daily lives; however, once they are put inside the robots, they become bigger and stronger.

Robots, Children, and Subjectivity

As we have seen, the robots and humans are connected in several ways within popular culture. The most obvious feature of robot anime is that all the central characters are children; toy companies even sponsor robot anime to sell plastic model robots to children. This practice is closely linked with postwar Japanese

15 For example, "'Mobile Suit Gundam' defined the current trajectory of Japanese robot anime. The detailed histories, mechanical images, realistic portrayals of humans, and references to abortive communication found in subsequent works in the genre all take Gundam as their standard." (Murakami 2005, pp. 144-145).

thought, in which the younger generation, who felt they were victims of the war, criticized the older generation for their wartime activity (Oguma 2002, p. 806). Anime creator Hayao Miyazaki made the following statement:

> "When we lost WWII, popular culture had to be changed. If the adults were central characters, that would be unforgivable. So after the war and until the *Phantom Detective* (Maboroshi Tantei, serialized in 'Shonen Manga' from 1957), the central characters were children. Was not Kaneda Shotaro, a child character in the anime *Tetsujin 28-go,* wiser than the chief of police? This logic was accepted not only by children, but by adults as well, who wrote that because children were innocent, they did not have responsibility for the defeat." (Miyazaki 2002, p. 20)

In the years following the defeat, the slogan of the intellectuals was to establish "subjectivity" and overthrow the "Emperor system."[16] Some people thought that the Showa emperor was a "robot" of the policymakers (Inoue 2007). At the same time, intellectuals started questioning the responsibility for the War (Oguma 2002, p. 104). Within these circumstances, adults could not be depicted as subjects. The defeat in WWII made a significant impression on Japanese culture in diverse ways. Popular artist Takashi Murakami states that postwar Japanese subculture has been carrying a scar of defeat; the atomic bombs and US occupation left Japanese people impotent. Murakami curated the exhibition Little Boy: The Arts of Japan´s Exploding Subculture in New York. This exhibition was a result of an invitation from the museum to the artist to illustrate his views on contemporary Japan. The director of the Japan Society Gallery Alexandra Monroe introduced this exhibition with these words:

> "To most Japanese, the term 'Little Boy' conjures up memories of catastrophic defeat and represents a narrative of national humiliation. To Murakami, its meaning and imagery also suggest the culture and politics of infantilization. The Japanese people, in his view, have developed a dependency on the US as 'protector' and 'superpower' that began with the American-led occupation (1945–1951) and that continues to this day, resulting in a willful negation of both adulthood and nationhood. In other words, the Japanese have refused – or rather, have been refused the chance – to grow up." (Monroe 2005, p. 246)

---

16 As for subjectivity in postwar Japan, see Koschmann 1996. The Japanese <Shutai-sei> is a vague concept that involves two meanings of "Subjectivity" and "Independence." This concerned with the traditional problem of "determinism and freedom" of human action, and from sociological perspective, offered against social constraint as one of the determinism. <Shutai-sei> as "Subjectivity" is found in A. Schutz´s action-theory. It is the autonomous function of "System of Relevance," that constitutes "World of everyday life" which is the cognitive base of action. (Kurebayashi 1988, p. 263).

In robot anime, the relationship between robots and children is very close. Children are weak and have undeveloped minds.[17] They become stronger through robots; therefore, the Japanese projected their identity into the characters of robot anime who were empowered by technology. In this way, the representation of robots shows us the problem of identity in postwar Japan.

The Japanese came to terms with the US′ political infiltration, as if they were robots. Under such circumstances, Japanese popular culture could not choose, but depicted robots as familiar; the "children" in postwar anime appear in this form of Japanese national identity. Children (the Japanese) who cannot (could not) be adults connect with robots and become strong. In other words, the problem of identity is resolved through technology. In 1985, Tezuka explained Astro Boy as follows:

> "Robots become a human enemy when they become aware of what they themselves are. The good robots that trust humans are the same as the good children who trust adults (obey their parents). However, the children who obey their parents might be weak and immature. Astro Boy is a parody of this. Astro Boy is Pinocchio. It gradually grows from immature to mature, incomplete to complete. Then at last, Astro Boy stands against the human. But Dr. Ochanomizu catch atom and repair it. Then Astro Boy becomes a friend of justice for the human. This is a great dilemma for me. When I tried to make Astro Boy human or greater than human, he only resists becoming human."[18]

Astro Boy is a coming-of-age story. Astro Boy is an analogy of both children and technology. He cannot become an adult no matter how many years pass.

Conclusion

The Japanese find their political and technological identity through robots. Because robots connect humans and technology, self and other, they emerged within the complex relationship of technology and identity. In prewar Japan, the notion of the robot signified a passive subjectivity; the word "robot" was used to represent those who had lost their political subjectivity and were controlled by someone else. In postwar Japan, "robot" became more familiar; people could more easily sympathize with robots because the Japanese had to think of robots positively. The Japanese completely lost their political subjectivity with their defeat in WWII. They became politically dependent on the American sphere of influence. Then the children in postwar anime expressed the Japanese problem of identity after WWII. The robot anime settled this previously unsolvable problem

17 Susan Napier writes, "The fact that these are children makes their vulnerability particularly disturbing, suggesting extratextual aspects of a social malaise in which young people seem less and less connected, not only with other people but also with themselves." (Bolton et. al. 2007, p. 119).

18 Interview with Osamu Tezuka. In Mainichi Chugakusei Shinbun, July 7, 1985. (Referred from Saito 2000, p. 30).

of Japanese subjectivity by depicting robots, which had lost their subjectivity, as positive entities. In this sense, robots, which have power and technological excellence, saved the Japanese identity. Technology (including national technological excellence) complements the loss of Japanese subjectivity after WWII. At one level or another, the historical context that this paper examined has been influenced and shaped by the unique robotics culture in Japan.

References

Bolton, Ch., I. Csicsery-Ronay (Jr.) & T. Tatsumi (eds.) 2007: Robot Ghosts and Wired Dreams: Japanese Science Fiction from Origins to Anime, Minneapolis & London.

Calichman, R. (ed.) 2008: Overcoming Modernity: Cultural Identity in Wartime Japan, New York.

Fukumoto, K. 2008: Karakuri Gijutsushibahnasi (rpt. in Collection of Kazuo Fukumoto Work vol.7), Tokyo.

Hirose, M. 2002: "The very first story of ASIMO by his 'founder' (Uminooya ga Kataru Asimo Tanjo Hiwa)." In: IT media news of December 4, 2002: http://www.itmedia.co.jp/news/0212/04/nj00_honda_asimo.html (accessed April 15, 2012).

Hornyak, T. N. 2006: Loving the Machine, Tokyo.

Inoe, H. 1993: Nohon Robotto Soseiki: 1920-1938, Tokyo.

Inoue, H. 2007: Nihon Robotto Sensoki, 1939-1945, Tokyo.

Ito, G. 2005: Dezuka is Dead: Towards an Expanded Discourse of Manga Expression, Tokyo.

Ito, K. 2003: "'Mr. F' and 'Astroboy': Images of Robots and Public Perceptions of Technology in before and after WWII." In: Japan Journal for Science, Technology & Society, vol. 12, 2003, pp. 39-63.

Ito, K. 2010a: "Vor Astro Boy." In: Technikgeschichte, 77(4), December 2010, pp. 353-372.

Ito, K. 2010b: "Astroboy´s birthday: Theorizing Science, Technology, and Popular Culture in Contemporary Japan." To be published in: East Asian STS Journal.

Ito, K. 2010c: "Robots, A-Bombs, and War." In: Jacobs R. (ed.) 2010: Filling the Hole in the Nuclear Future, Lexington Books, pp. 63-97.

Koschmann, J.V. 1996: Revolution and Subjectivity in Postwar Japan, Chicago.

Kurebayashi, N. 1988: "Reflections on the Concept of <Shitai-sei> - Focusing on 'Subjectivity' and 'Independence' of Action -." In: Bulletin of the Faculty of Education, University of Tokyo, vol. 28, pp. 253-271.

MacDorman, K. F., S. K. Vasudevan & Ch.-Ch. Ho 2009: "Does Japan really have robot mania? Comparing attitudes by implicit and explicit measures." In: AI & Soc. (2009) 23, pp. 485–510.

Miyazaki, H. 2002: "40,000 words interview with Hayao Miyazaki." In: Sight, vol. 10, winter 2002 winter, pp. 8-43.

Mizuno, H. 2009: Science for the Empire: Scientific Nationalism in Modern Japan, Stanford (Calif.).

Monroe, A. 2005: "Introducing Little Boy." In: Murakami, T. (ed.) 2005: Little Boy: The Arts of Japan´s Exploding Subculture, New York, pp. 241-261.

Murakami, T. (ed.) 2005: Little Boy: The Arts of Japan´s Exploding Subculture, New York.

Napier, S. J. 2005: Anime from Akira to Howl´s moving castle: Experiencing contemporary Japanese animation, New York.

Oguma, E. 2002: Minshu to Aikoku (Democracy and Patriotism), Tokyo.

Richter, St. 2010: "Sign of 'Cool Japanology' in Europe." In: Azuma H. (ed.) 2010: Future of Japanese Imagination: Possibility of Cool Japanology, Tokyo, pp. 169-185.

Saito, T. 2000: Sento Bishojo no Seishinbunseki (Psychoanalysis of Beautiful Fighting Girl), Tokyo.

Shimotsuki, T. (ed.) 1997: Tanjo! Tezuka Osamu (Birth Tezuka Osamu), Tokyo.

Suzuki, J. 2010: Kagaku Gijutsu Seisaku, Tokyo.

Tanabe, H. 1936: Kagaku Seisaku no Mujun, Kaizo.

Tezuka, O. 1997: Boku no Manga Jinsei, Tokyo.

Yamaguchi, A. 1990: Norakuro wa Warera no Dojidaijin (Norakuro is Our Coeval), Tokyo.

Yoshida, M. (ed.) 2003: Ninshin suru Robot: 1920 nendai no Kikai to Shintai (Fall Pregnant Robot: Machine and Body of 1920s), Kanagawa.

Acknowledgements

This study is supported by JSPS, KAKENHI (Grants-in-Aid for Scientific Research, NO. 21520004) and Global COE Program The University of Tokyo Center for Philosophy (UTCP) from MEXT, Japan.

Understanding the Feasibility and Applicability of the Musician-Humanoid Interaction Research: A Study of the Impression of the Musical Interaction

Jorge Solis & Atsuo Takanishi

Introduction

The recent technological advances in robot technology, musical information retrieval, artificial intelligence, etc. are enabling humanoid robots to roughly emulate the physical and perceptual capabilities of musicians while playing musical instruments. In particular, a wind instrument playing humanoid robot requires many different complex systems to work together integrating musical representation, techniques, expressions, detailed control and sensitive multimodal interactions within the context of a piece, as well as interactions between performers. Due to the inherent interdisciplinary nature of the topic, this research can contribute to the further enhance musical understanding, interpretation, performance, education and enjoyment. In this Chapter, an overview of the research on the development of an anthropomorphic flutist robot in order to enable it to interact with human musicians is given. A set of quantitative and qualitative experimental evaluations are performed to understand the real possibilities to enable its interaction with human musicians. Furthermore, the impressions of the interaction from the point of view of the human musicians are presented and discussed.

The development of wind instrument playing humanoid robots has interested the researchers since the golden era of automata up to today. As an example, we may find some classic examples of automata displaying human-like motor dexterities to play instruments such as the "Flute Player" (Vaucanson 1979). In addition, we find the first attempt to develop an anthropomorphic musical robot, the WABOT-2. The WABOT-2 was capable of playing a concert organ, built by the late Prof. Ichiro Kato. In particular, Prof. Kato argued that the artistic activity such as playing a keyboard instrument would require human-like intelligence and dexterity (Kato et al. 1987). Compared to other kinds of instruments (i.e. piano, violin, etc.), the research on wind instruments have interested researchers from the point of view of human science (i.e. study of breathe mechanism), motor learning control (i.e. coordination and synchronization of several degrees-of-freedom), musical engineering (i.e. modeling of sound production.), etc.

During the last decades, the field of Humanoid Robotics and its application to Human Robot Interaction (HRI) continues to grow around the world. The research on human-robot interaction (HRI) has been an emerging topic of interest for both basic research and customer application. The current studies focus on behavioral and cognitive aspects of the interaction and the social contexts surrounding it. HRI issues have long been a part of robotics research because the goal of fully autonomous capability has not been met yet. From the scientific point of view, one of the most challenging problems is giving the robots an understanding of how to interact with human beings at the same logical level. It is therefore expected that robots may act as active agents that can drive the human interaction, instead of merely reproducing a sequence of motions.

In addition, Humanoid Robots are also used for the validation of bioengineering models for increasing the knowledge of biological subsystems. The idea is to develop a physical model at both levels: at the level of sensors and actuators and at the level of sensory data processing, sensory-motor coordination and behavioral schemes (Asada et al. 2001; Rucci et al. 1997; Scassellati 2002). From the application point of view, some companies have stated that the relationship between robots and humans is a factor that designers have to explore more deeply for the successful integration of the HR into society (Ishida et al. 2001; Toyota Robot Partner; etc.). The integration of HR into society will trigger widespread social and economic changes, for which public and private policies must now be contemplated (Capurro 2000). Here, personal interactions might prompt or cause humans to form unidirectional emotional attachment relationships with robots that are not only appropriately reciprocated, but might allow robots to take advantage of people´s emotional propensities and reactions (Scheutz 2009). Actually, many roboticists, as well as authoritative scholars of the history of science and technology, have already labeled the 21st Century as the *age of the robots* (Brooks 2002).

In recent years as an initiative to promote the discussion among robotics, the Technical Committee on Roboethics has been introduced (Veruggio & Operto 2008). Roboethics is an applied ethics whose objective is to develop scientific/cultural/technical tools that can be shared by different social groups and beliefs. Even that several researchers from different fields (i.e. robotics, physiology, ethics, law, etc.) have started to discuss the possible impacts of the introduction of HRs (i.e. robot-addiction social problem, etc.); the lack of knowledge (from the society regarding the current state-of-art in the development of humanoid robotic) complicates the understanding of Roboethics. In this Chapter, the development of wind playing instrument humanoid robots is introduced. Then, an overview of the development of an anthropomorphic flutist robot is given by stressing their capabilities to interact with musicians. Finally, some possible implications are point out.

The authors; in a previous publication (Solis & Takanishi 2010), have introduced the recent research trends on the field of Humanoid Robotics, their possible applications (e.g. medical field, etc.) and their possible impact were discussed. In this chapter, a more detailed overview of the recent research achievements on the development of musical robots and their possible implications are introduced.

Wind Playing Instrument Humanoid Robots

During the golden era of automata, the "Flute Player" developed by Jacques de Vaucanson was designed and constructed as a means to understand the human breathing mechanism. Vaucanson presented "The Flute Player" to the Academy of Science in 1738. For this occasion he wrote a lengthy report carefully describing how his flutist can play exactly like a human. The design principle was that every single mechanism corresponded to every muscle (Vaucanson 1979). Thus, Vaucanson had arrived at those sounds by mimicking the very means by which a man would make them. More recently, the "Flute Playing Machine" developed by Martin Riches was designed to play a specially-made flute somewhat in the manner of a pianola, except that all the working parts are clearly visible (Sadowsky 2005). The Flute Playing Machine is composed of an alto flute, blower (lungs), electro-magnets (fingers) and electronics. The design principle is basically transparent in a double sense. The visual scores can be easily followed so that the visual and acoustic information is synchronized. Other example is the Waseda Flutist Robot No.4 Refined IV (WF-4RIV) that has been developed by the authors (Solis et al. 2008). This research is focused on understanding the human motor control from an engineering point of view, understanding how to facilitate the interaction between the robot and humans, and proposing novel ways of entertainment. The WF-4RIV has a total of 41-DOFs and it is composed of the following simulated organs (Solis et al. 2008): lungs, lips, tonguing, vocal cord, fingers, and other simulated organs to hold the flute (i.e. neck and arms).

On the other hand, one of the first attempts to develop a saxophone-playing robot was done by Takashima at Hosei University (Takashima & Miyawaki, 2006). His robot; named APR-SX2, is composed of three main components: mouth mechanism (as a pressure controlled oscillating valve), the air supply mechanism (as a source of energy), and fingers (to make the column of air in the instrument shorter or longer). The artificial mouth consisted of flexible artificial lips and a reed pressing mechanism. The artificial lips were made of a rubber balloon filled with silicon oil with the proper viscosity. The air supplying system (lungs) consists of an air pump and a diffuser tank with a pressure control system. The APR-SX2 was designed under the principle that the instrument played by the robot should not be changed. A total of twenty-three fingers were configured to play the saxophone´s keys (actuated by solenoids), and a modified mouth mechanism was designed to attach it to the mouthpiece, no tonguing mechanism was implemented (normally reproduced by the tongue motion). Other example is the Waseda Saxophonist Robot No.2 Refined (WAS-2R) that has been developed by the authors (Solis et al 2011). This research is focused on enabling the interaction with musical partners (i.e. with the WF-4RIV). Therefore; as a long-term goal, we expect that the proposed saxophonist robot is able not only of performing a melody, but also to dynamically interact with the musical partner (i.e. walking while playing the instrument, etc.). The WAS-2R is composed by 22-DOFs that reproduce the physiology and anatomy of the organs involved during the saxophone playing as follows (Solis et al. 2011): 3-DOFs to control the shape of the artificial lips, 16-DOFs for the human-like hand, 1-DOF for the tonguing mechanism and 2-DOFs for the lung system.

Humanoid-Musician Interaction with the WF-4RIV

Conventionally, the humanoid music robots are mainly equipped with sensors that allow them to acquire information about their environment. Based on the anthropomorphic design of humanoid robots, it is therefore important to emulate two of the human´s most important perceptual organs: the eyes and the ears. For this purpose, the humanoid robot integrates in its head, vision sensors (i.e. CCD cameras) and aural sensors (i.e. microphones) attached to the sides for stereo-acoustic perception. In the case of a musical interaction, a major part of the typical performance (i.e. Jazz) is based on improvisation. In these parts musicians take turns in playing solos based on the harmonies and rhythmical structure of the piece. Upon finishing his solo section, one musician will give a visual signal, a motion of the body or his instrument, to designate the next soloist. Another situation of the musical interaction between musicians is basically where the higher skilled musician has to adjust his/her own performance to the less skilled one. After both musicians get used to each other, they may musically interact toward enabling the multimodal interaction between the musician and humanoid robot, the Musical-based Interaction System (MbIS) has been proposed. The MbIS has been conceived for enabling the interaction between the musical robot and musicians. The proposed MbIS is composed by two levels of interaction that enables partners with different musical skill levels to interact with the musical robot (Petersen et al. 2010a): intermediate-based and advanced-based level of interaction. The purpose of the two-level design is to make the system usable for people with different experience levels.

In the basic interaction level we focus on enabling a user who does not have much experience in communicating with the robot to understand about the device´s physical limitations. We use a simple visual controller (implemented by virtual button and faders) that has a fixed correlation regarding which performance parameter of the robot it modulates, in order to make this level suitable for beginner players. The WF-4RV is built with the intention of emulating the parts of the human body that are necessary to play the flute. Therefore it has artificial lungs with a limited volume. Also other sound modulation parameters like the vibrato frequency (generated by an artificial vocal cord) have a certain dynamic range in which they operate. With the extended level interaction interface, our goal is to give the user the possibility to interact with the robot more freely. To achieve this, we propose a teaching-in system that allows the user to link instrument gestures with musical patterns. Here, the correlation of sensor input to sensor output is not fixed. Furthermore, we allow for more degrees-of-freedom in the instrument movements of the user. As a result this level is more suitable for advanced level players. We use a particle filter-based instrument gesture detection system and histogram-based melody detection algorithms (Petersen et. al 2010b). In a teaching phase the musician can therefore assign instrument gestures to certain melody patterns. In a performance phase the robot will re-play these melodies according to the taught-in information.

What kind of impressions musicians have about interacting with the WF-4RIV?

In order to understand the impression of musicians while interacting with the WF-4RIV, a set of experiments were carried out. In particular, experiments were carried out to evaluate how well a musical partner can express his musical intentions using the proposed two stage mapping approach. In case of the beginner level interaction interface experiment, the robot is controlled by one virtual fader. This fader is used to continuously control the speed of a pre-defined sequence that is played by the flutist robot. The output of the sensor processing system determining the value of the virtual fader is conditioned by the lung movement of the robot.

In order to perform the experiment, a professional flutist player is situated in front of the robot. After introducing the functionality of the beginner level stage to the player we recorded data of the resulting interaction with the robot. To achieve quantitative results for the first level interaction system we performed the experiment with a professional flutist player. For the proposed experiments, the fader movements control the tempo of the tone sequence that is performed by the robot. If the amount of air remaining in the lung reaches a certain limit, the fader value transmitted to the robot is faded-out.

In the experiment for the advanced level interaction system, two phases have been defined: the teaching phase and the performance phase. In the first phase the interacting musician teaches a movement-performance parameter relationship to the robot. In this particular case we relate one of three melody patterns to the inclination angle of the instrument of the robot´s partner musician. From this information the robot builds a table that relates instrument angles to musical patterns. In the second stage the interaction partner controls the robot with these movements.

The transition of the teaching phase to the performance phase is defined by the number of melody patterns associated by the robot. In case of this experiment, the switch occurs after 3 melody patterns have been recorded. After introducing the functionality of the system to the player, he performed one teaching phase and the following performance phase. The results for the advanced level interaction experiment, during the teaching phase, the musician related three single notes A4, B4 and C5, to angles of 110, 93 and 60 degrees. The robot switches from the teaching phase to the performance phase after three notes /

note patterns have been recorded. In the performance phase the musician varies the inclination angle of the flute. With the inclination changing, also the note played by the robot changes.

To provide qualitative results documenting the usability of the system we performed the described experiments with two beginner-level, two intermediate-level and two professional level instrument players. We investigated their impression of the interaction quality with a questionnaire. This questionnaire asked the experiment subject to evaluate the system in three categories on a scale from 1 (insufficient) to 10 (excellent). The three categories to be questioned were proposed as follows (Petersen et. al 2010b):

- *Overall Responsiveness of the System*: In order to find out how the subjects responded to the technical implementation of the system in terms of detection and processing speed,
- *Adaptability to Own Skill-Level*: In order to find out in how far the separation of the system in beginner level and advanced level interaction system fits for the differently experienced players.
- *Musical Applicability / Creative Inspiration*: In order to enquire about how the musicians felt they could express their musical intentions through utilizing the interaction interface.

From the experimental results obtained in the survey, the *Overall Responsiveness of the system*, we may observe higher grades for the less experienced players and lower grades for the experienced players. With higher skill level the requirement for responsiveness seem to increase. On the other hand, the *Adaptability to Own Skill Level* we observed that according to our expectations the less experienced players would feel more comfortable with the beginner level interaction system and the more experienced players would give higher grades in case of the advanced level interaction system. Finally, as for the *Musical Applicability / Creative Inspiration*, the experimental results show intermediate scores for all skill levels.

From the above the results, we (roboticists) are focusing in embedding motor dexterities and cognitive capabilities into humanoid robots towards enabling a natural interaction between humans and robots. For this purpose, biologically-inspiration for the mechanical design and control as well as advanced techniques for signal processing have been implemented into the flutist robot. Even though it is evident that musicians are required to have certain experience to effectively interact with the flutist robot, it seems that their expectations are increased (depending on their skills) so that they would feel the interaction could gradually become more natural (of course, still several technical issues; i.e. body gestures, etc., should be considerably improved). As a result, even the technical challenges we (roboticists) may still have as further development, they (musicians) would consider the possibility to tightly interact with robot as an approach to find novel ways of musical expression. In other words, from the point of view of engineering, we may contribute to understand better the mechanism that facilitates the communication among humans in musical terms by embedding motor dexterities and cognitive capabilities into the flutist robot. Furthermore, from the point of view of entertainment, even though the musicians who participated in the proposed experiment consider that the robot would be able to interact with them with certain limitations, they still consider the robot could be used as an approach for creative inspiration. Therefore, they (musicians) may find rather preferable interact with humanoid robots rather than human partners while creating new ways of expression.

Conclusions

In this Chapter, an overview of the research on the development of wind playing instrument humanoid robots has been introduced. In particular, the development of the Waseda Flutist Robot No. 4 Refined IV (WF-4RIV) was detailed. An overview of the proposed Musical-based Interaction System (MbIS) was given. In order to understand how well a musician can express his/her musical intentions while interacting with the flutist robot. Furthermore, a survey of the impressions from musicians with different levels of expertise was carried out to understand their impressions while interacting with the proposed system.

From the experimental results; the flutist robot was able to effectively process the incoming information from the musical partner while still the physical limitations of the robot were considered (i.e. amount of air remaining in the lung mechanism). Furthermore, the flutist robot was able to learn the musical intentions of the musician after the teaching phase is completed. Even though a limited number of notes were considered, the intentions of the musician were reproduced by the flutist robot. On the other hand; from the results obtained from survey, we may notice that the possibility of creating a natural musical interaction between the human and the robot also depends on the level of expertise and the time the musician expend to learn the interactive capabilities of the robot.

Therefore; we should consider than even several technical challenges to solve and the musical partner should obtain sufficient experience to interact with the robot, the musicians consider the robot could be used as a new musical tool for inspiration. However, there is no detailed study about the possible impact may have the introduction of musical robots on the musical field. Would be possible to face a risk that musicians will rather prefer robots instead of humans to as a source of musical inspiration? Due to the scarcely research done in order to understand the possible impact in the future, the authors believe such issues should be discussed by involving researchers from different fields (in fact, this is one of the priorities of the IEEE/RAS Technical Committee on Roboethics.) such as engineering, social science, music, etc.

References

Asada M., K. F. MacDorman, H. Ishiguro & Y. Kuniyoshi 2001: "Cognitive developmental robotics as a new paradigm for the design of humanoid robots." In: Robotics and Autonomous Systems, 37, pp. 185-193.

Brooks, R. 2002: Flesh and Machines. How robots will change us, New York.

Capurro, R. 2000: "Ethical Challenges of the Information Society in the 21st Century." In: International Information & Library Review, 32, pp. 257-276.

Ishida, T., Y. Kuroki, J. Yamaguchi, M. Fujita & T. Doi 2001: "Motion entertainment by a small humanoid robot based on OPEN-R." In: Proc. of the 2001 IEEE/RSJ International Conference on Intelligent Robots and Systems, 2, pp. 1079-1086.

Kato, I., S. Ohteru, K. Shirai, T. Matsushima, S. Narita, S. Sugano, T. Kobayashi, & E. Fujisawa 1987: "The robot musician 'WABOT-2' (waseda robot-2)." In: Robotics 3 (2), pp. 143-155. (Available at: http://www.sciencedirect.com/science/article/pii/0167849387900027 [accessed June, 8, 2012].)

Petersen, K., J. Solis & A. Takanishi 2010a: "Musical-Based Interaction System for the Waseda Flutist Robot: Implementation of the Visual Tracking Interaction Module." In: Autonomous Robots Journal, Vol. 28 (4), pp. 439-455.

Petersen, K., J. Solis & A. Takanishi 2010b: "Implementation of a Musical Performance Interaction System for the Waseda Flutist Robot: Combining Visual and Acoustic Sensor Input based on Sequential Bayesian Filterings." In: Proc. of the IEEE/RSJ International Conference on Intelligent Robots and Systems, pp. 2283-2288.

Rucci, M., J. Wray, G. Tononi & G. M. Edelman 1997: "A robotic system emulating the adaptive orienting behavior of the barn owl." In: Proc. 1997 IEEE International Conference on Robotics and Automation, 1, pp. 443-448.

Sadowsky, Th. 2005: Martin Riches - Maskinerne / The Machines. Heidelberg.

Scassellati, B. 2002: "Theory of mind for a humanoid robot." In: Autonomous Robots, 12 (1), pp. 13-24.

Scheutz, M. 2009: "The Inherent Dangers of Unidirectional Emotional Bonds between Humans and Social Robots." In: Proc. of the ICRA2009 Workshop on Roboethics.

Solis, J., K. Taniguchi, T. Ninomiya & A. Takanishi 2008: "Understanding the Mechanisms of the Human Motor Control by Imitating Flute Playing with the Waseda Flutist Robot WF-4RIV." In: Mechanism and Machine Theory, 44 (3), pp. 527-540.

Solis, J. & A. Takanishi 2010: "Recent trends in humanoid robotics research: scientific background, applications and implications." In: Accountability in Research, Vol. 17, pp. 278-298.

Solis, J., M. Takeuchi, K. Takafumi, K. Petersen, S. Ishikawa, A. Takanishi & K. Hashimoto 2011: "Improvement of the Oral Cavity and Finger Mechanisms and Implementation of a Pressure-Pitch Control System for the Waseda Saxophonist Robot." In: Proc. of the International Conference on Robotics and Automation.

Takashima S. & T. Miyawaki 2006: "Control of an automatic performance robot of saxophone: performance control using standard MIDI files." In: Proc. of the IEEE/RSJ Int. Conference on Intelligent Robots and Systems - Workshop: Musical Performance Robots and Its Applications, pp. 30-35.

Toyota Robot Partner: http://www.toyota-global.com/innovation/partner_robot/ (accessed March 11, 2013)

Vaucanson, J. de 1979: Le Mécanisme du Fluteur Automate. An Account of the Mechanism of an Automation or Image Playing on the German-Flute, Buren.

Veruggio G. & F. Operto 2008: "Roboethics: Social and Ethical Implications of Robotics." In: Siciliano, B. & O. Kathib (eds.) 2008: Springer Handbook of Robotics, Berlin, pp. 1499-1524.

Acknowledgment

A part of the research on Musical Instrument-Playing Robot was done at the Humanoid Robotics Institute (HRI), Waseda University and at the Center for Advanced Biomedical Sciences (TWINs). This research is supported (in part) by a Gifu-in-Aid for the WABOT-HOUSE Project by Gifu Prefecture. This work is also supported in part by Global COE Program "Global Robot Academia" from the Ministry of Education, Culture, Sports, Science and Technology of Japan. WF-4RIV and WAS-2R have been designed by 3D CAD design software SolidWorks. Special thanks to SolidWorks Japan K.K. for the software contribution. The study on the Waseda Saxophonist Robot is supported (in part) by a Grant-in-Aid for Young Scientists (B) provided by the Japanese Ministry of Education, Culture, Sports, Science and Technology, No. 23700238 (J. Solis, PI). Finally, this study is in part supported by JSPS, KAKENHI (Grants-in-Aid for Scientific Research), No. 21520004 (K. Ishihara, PI).

Mozart to Robot – Cultural Challenges of Musical Instruments

Michael Funk & Jörg Jewanski

In this paper, emphasis is not on the enormous technical richness of mechanical musical instruments or music automates. Rather, the focus is on the cultural and social implications of new musical instruments: 1. What are the reasons for composers to develop musical pieces for mechanical instruments (like the musical clock or the player piano in the 18th- and 20th-centuries)? What are the characteristics of this music? 2. How could musical robots challenge composers, musicians and educators today?

The Composer's View on Mechanical Instruments as seen in the Music of Wolfgang Amadeus Mozart and Conlon Nancarrow

With regard to mechanical musical instruments,[1] Wolfgang Amadeus Mozart (1756-1791) and Conlon Nancarrow (1912-1997) can be seen as two chronologically and aesthetically representative composers. With respect to their musical work, two important instruments will be introduced in this section: the musical clock and the player piano. Both became popular in the 18th- and 19th-centuries and trace back to mechanical instruments like artificial singing birds, and automates with autonomous pipe organs – controlled by barrels and driven by water –, which had been developed in pre-Christian times.

The most important compositions for clock watches have been written by Mozart. For this instrument, a mechanical clockwork activates a barrel that causes pipe sounds at regular intervals, e.g. every full hour. Mozart wrote his compositions on traditional music paper, the transfer to the barrel was realized by someone else. The *Köchel catalog* lists five musical works for these instruments:

1. *Adagio für (Kl. od.) eine Orgelwalze*, KV 593a [Fragment of 9 bars, probably a draft to KV 594, undated]
2. *Ein Stück für ein Orgelwerk in einer Uhr* (December 1790), KV 594
3. *Ein Orgel Stücke für eine Uhr* (March 1791), KV 608
4. Andante, KV 615a [Fragment of 4 bars, probably a draft to KV 616, undated]
5. *Ein Andante für eine kleine Walze in eine kleine Orgel* (May 1791), KV 616

1 An overview of mechanical music instruments can be found in: Buchner 1992; Hocker 1996 (with huge bibliography); Ord-Hume 2001.

Mozart composed this music for financial reasons, as he wrote to his wife in the following letter (Oct. 3rd 1790):

> "I have now made up my mind to compose at once the Adagio for the watchmaker and then to slip a few ducats into the hand of my dear little wife. And this I have done; but as it is a kind of composition which I detest, I have unfortunately not been able to finish it. I compose a bit of it every day – but I have to break off now and then, as I get bored. [...] If it were for a large instrument and the work would sound like an organ piece, then I might get some fun out of it. But, as it is, the works consist solely of little pipes, which sound too high-pitched and too childish for my taste." (Anderson 1938, pp. 1403-1404)

Tonal reasons spoiled Mozart´s enjoyment of composing mechanical music. Anyway, his wife got an audition with the count Joseph Deym von Stritez, who was planning a crypt with a coffin out of glass for a recently deceased field marshal. Because of the graceful atmosphere in the crypt, the count ordered a piece of funeral music for the mechanical clock from Mozart (Plath 1982, p. XXI). After Mozart´s intervention and because of the tonal reasons, the original small instrument was replaced by a bigger one.

No doubt, the music is of high quality: the Adagio KV 594 is characterized by an ABA-form with falling chromatic lines in F minor and motives of sighs in the beginning and end (evidently in the sense of funeral music), while the middle part in F major is shaped by fanfares and several motives in winged tempo. After this lively middle part, the contrast to the end part becomes stronger and the feeling of sorrow is enhanced. KV 608 with its fugato in four voices, many pseudo entries after the exposition and a modulation from F minor to F# minor became well known also as transcription for piano (four hands). Even Ludwig van Beethoven was drawing up his own copy of it (Dittrich 2005, pp. 555-557).

Nancarrow[2] also wrote compositions on traditional music paper first but also transferred them personally to piano rolls in a time consuming process. The piano roll belongs to the player piano, a self-playing piano instrument, which is able to produce extreme rates and complexities of sound. No human pianist, not even the greatest virtuoso, could realize such fast music. A human pianist is able to play virtuous tone scales of ca. 15 tones per second, a player piano can reach up to 100. The number of simultaneous tones on the player piano is also exceeding the bodily and sensorimotor possibilities of human hands. Those tones also do not need to belong to an ambit of tenth, which is the range of a human hand.

Because of three interrelated reasons, Nancarrow was composing for the player piano: First, already in his early traditional instrumental music, rhythm, meter and tempo started becoming more important for him than other parameters. His works in the 1930s and early 1940s are characterized by very high tempo and complex elaborated rhythm. The player piano was a technical opportuni-

2 The next passages are following the descriptions in Hocker 2002; Fürst-Heidtmann 2004; Herzfeld 2007.

ty to gain and follow these principles of composition in an uncompromising way, without taking care of the bodily capacities and limits of human musicians. For Nancarrow, the player piano provided nearly no limit to tempo and complex rhythms. The second reason was his dissatisfaction with performers, who did not work out the characteristics of his music adequately. The third reason resulted from political circumstances and Nancarrow´s artistic retreat from New York to Mexico City in 1940, where he lived till the end of his life and where he also had less contact to virtuous pianists.

After this, between 1949 and 1993, his main compositions (ca. 50, depending on the way of counting) had been developed as: "Studies for Player Piano." This work is shaped by experimental musical stratification of different tempo-layers: In NR 21, the high and low voice are speeded up and slowed down in reversed tempo. In NR 27, one voice provides a constant ostinato, while four others bear different tempo variations. In the late studies, dynamic glissandi emerge with fast sequences of trills or arpeggios upon the whole claviature (NR 40, 41 and 48), but also enormous complex pieces like a canon of twelve voices (NR 37). The aesthetic impression of this music is dominated by virtuous effects, but Nancarrow also used symmetric structures, proportions and number-relations while he even included elements of Jazz and Ragtime in his compositions.

Mozart and Nancarrow – two composers with two different approaches to mechanical instruments – stand in a significant tradition. Other major European composers in the 18th- and early 19th-century have developed music for musical watches (and similar instruments): Carl Philipp Emanuel Bach, Wilhelm Friedemann Bach (for a long time his music was mistaken for compositions of Johann Sebastian Bach), George Frideric Handel, Joseph Hadyn, Ludwig van Beethoven and Luigi Cherubini. Nancarrow had ancestors as well: Igor Stravinsky, Paul Hindemith, George Antheil and Henry Cowell, in whose tradition Nancarrow was working. In the following section, the focus is on current challenges of musical robots and some of their aesthetic and pedagogical implications and potentialities. What can we learn from Mozart, Nancarrow and also from Beethoven for the usage of musical robots today?

Musical Robots as Cultural Challenge for Musicians, Composers and Pedagogues

Academic interest in musical robots has recently been documented in the study "Musical Robots and Interactive Multimodal Systems" (Solis & Ng (eds.) 2011). Here one basic focus is not on musical scores or the biography of individual composers, but on gesture-studies (Solis & Ng 2011, p. 2) and "the role of physical gestures in human-machine interaction" that is related to enaction, embodied cognition and generally the notion of action in music(s) (Bevilacqua et al. 2011, p. 127). Musical gestures can be understood as "meaningful combination[s] of sound and movement" (Godøy & Leman 2010, p. ix). Studying gestures in general means interdisciplinary research.

> "Consequently, the field has attracted researchers from a number of different disciplines such as anthropology, cognitive science, communication, neuroscience, psycholinguistics, primatology, psychology, robotics, sociology and semiotics, and the number of modern gesture studies has grown." (Ishino & Stam 2011, p. 3)

The same can be said for musical gestures (Gritten & King 2011, pp. 1ff.; Jensenius et al. 2010, p. 28; Leman & Godøy 2010, p. 10).

> "In the so-called embodied view of perception and cognition [...] motor schemas are seen as basic for all cognition, not only auditory perception. This means that all perception and reasoning, even rather abstract thinking, is understood as related to images of action." (Godøy 2011, p. 15)

This musicological statement correlates with current investigations in the cognitive sciences, where human cognition is seen as describable in terms of interaction with concrete social and technical environments (Noë 2004; Varela, Thompson & Rosch 1997), but also to philosophical theories of the human embodied mind and "*Leiblichkeit*" (Irrgang 2007; Irrgang 2009). The basis for human cognition, also in music, is body-movement. If we want to understand cultural and social implications of musical robots, then we need to understand our underlying meaningful human movements that are related to sound with or without robots. Musical gestures have a sensorimotor and a technological dimension as well. For Don Ihde, new musical instruments cause new embodied relations and technical mediations.

> "By this [embodied relations] I mean that the human or humans producing the music, do so through material artifacts or instruments. [...] Going into this practice, of course, there can be a learning, the development of special techniques, higher and higher skills of sound and music making, the development of styles, schools of musical traditions and instrumental developments. [...] Early instruments tended to be fairly simple even if widely varied, and I want to say, such simpler instruments also tended to demand highly skilled bodily movement." (Ihde 2007, p. 255)

Recording-technologies, amplifiers or processes of digitalization are some major aspects of new cultural and musical applications since the 20th-century (for more details: Ihde 2007, pp. 227-264 and also Leman 2008, pp. 137-184). Music robots could be the next step at the beginning of the 21st century. Robots will not replace musicians or composers, but they could provide a new perspective from which humans view themselves and their cultural environment within technically mediated social interactions. Robots are still tools, but tools that could shape "the development of special techniques, higher and higher skills of sound and music making, the development of styles, schools of musical traditions." In this context, Kia Ng differentiates between three categories of new musical interfaces: "Imitation of acoustic instruments," "Augmented Instruments" and "Alternative Controllers" (Ng 2011, pp. 107-108). The third category is related to "new interfaces that are not based on traditional instrumental paradigms […] [and] are original controllers which require learning new skills" (Ng 2011, p. 108). The aspect of learning new skills raises pedagogical questions. But there are also misunderstandings between classical music teachers and people that are open-minded to use new instruments like robots. Kia Ng summarizes this from the perspective of musical education:

> "Pedagogical applications are rare and of experimental nature, which can be explained partly by the community gap between music practitioners using such technologies and traditional music teachers." (Ng 2011, p. 108)

Musical robots also cause challenges for musical education, but they cannot replace music teachers. They still remain tools, but they may become an enriching addition as did keyboards or CD-players, which can be found in nearly every class room or school today.

The basic point, for musicians as well as for pedagogues, students or composers, is that human knowing is primarily shaped by tacit and sensory remembering, which includes movement, perception and its interpretation. Godøy argues that

> "projecting images of sound-action relationships from past musical and environmental sonic experiences onto new musical instruments could be seen as a case of anthropomorphic, know-to-unknown, projection, and as a matter of basic functioning of our mental apparatus, what we see as a motormimetic element in music perception […]." (Godøy 2011, p. 15)

We interpret new instruments and sounds always in the horizon of movements and perceptions that we have already been used to. So our experiences with common instruments shape the way in which we start to develop new skills, sounds or musical aesthetic. As already indicated in the first section, Conlon Nancarrow is one example for that, because his first steps in composition he had done in terms of classical composition work for human instrumentalists. In the horizon of these experiences and the related pre-knowledge, he started his main work oriented towards the player piano in the early 1940s. Today, the same could happen with musical robots. On the one hand, these new instruments can challenge composers aesthetically, or at least they may provide some pragmatic jobs (as was the case for Mozart with regard to his works for the musical watch). On the other hand, and here we evidently come close to the basic aspect of motormimetic interaction, music robots can challenge musicians that are directly technically interacting with those instruments.[3] But once again, those robots cannot replace human knowing of musicians, educators or composers.[4]

3 See also the chapter "Understanding the Feasibility and Applicability of the Musician-Humanoid Interaction Research: A Study of the Impression of the Musical Interaction" (Jorge Solis & Atsuo Takaneshi) in this book.

4 See also the chapter "Humanoid Robots and Human Knowing" (Michael Funk) in this book.

Beneath the surface of those observations lurks a philosophical hypothesis: musical aesthetic is not only a matter of musical scores and harmonic theories, but moreover also a question of concrete technical praxis that also depends on the possibilities of musical instruments.[5] In this context we may understand musical robots as useful tools for music education, the development of new styles or musical (youth) cultures. In the end what happened to electronic music, the first synthesizers and so on could repeat itself. Musical robots can also cause new bodily, motor-mimetic and sensory perspectives for human gestural expressions. Creativity in this case does not depend on the robot, but on what humans are starting to do with it.

> "Nevertheless, social interaction still is one of the most important factors in music performance (e.g. interaction between performers, between performers and conductor, between performers and audience)." (Camurri & Volpe 2011, p. 61)

Music robots cannot replace emotional social interaction. But they can be part of a new way in which we start reading and expressing our social emotions and the related musical aesthetic. In the end, this is not really something new. With respect to Ludwig van Beethoven and his challenge in handling the limited claviature of his pianos we can see that technical potentials, but also technical limits can shape aesthetic expression.

Beethoven and the Limited Claviature

In the history of European classical music(s) many examples can be found for the many ways in which musical instruments can shape aesthetic aspects of compositions and musical scores. One prominent example seems to be Johann Sebastian Bach and "The Well-Tempered Clavier." But also by reference to Ludwig van Beethoven´s piano sonatas, the relation between the technical equipment of pianos and the compositions can be demonstrated. He had to take care of the range of the instruments at his time: the ambitus of the claviature and the low quality of the highest tones often constrained divergences from his (probably) original intentions (Bruckmann 1993, pp. 10-13). According to Beethoven´s friends it was for this reason that he composed many piano works in direct technical and bodily-sensory interaction while he was improvising with the instrument, and not while he was sitting away from the piano or behind an isolated desk (Bruckmann 1993, p. 13). From a philosophical and aesthetic point of view, it is amazing to realize that those limits of former pianos did not only cause second best solutions but also tonal improvement as it can be illustrated with respect to Opus 31, No. 2, 1. Movement. While in bars 59-62 Beethoven developed the descant in octaves, for the corresponding part in bars 189-192, he was forced by the limitation of the keyboard to remain in the position and

5 For a more detailed overview with respect to the current discussions in musical sciences and philosophy see also Funk & Coeckelbergh 2013.

achieve a gradation by harmonic means, which is even more intense than developing in octaves (Bruckmann 1993, p. 49).

What does this say about musical robots? If robots are not able to reproduce all intentions of composers or musicians that are interacting with them, this could lead to new unintended enhancement of musical sounds and aesthetic meanings. This depends from the creativity of humans, not from the robots themselves. In the horizons of musical traditions, educations and bodily pre-knowledge, Beethoven made something out of the piano; the piano was limited but did not tell him what he should do. With musical robots it is just the same.

Conclusion

Composing for the musical clock was more a pragmatic need and a banal job for Mozart rather than a great musical fruition. Besides the musical clock the player piano also has a prominent stand in the history of pre-robotic mechanical instruments. In the 20th-century, Nancarrow developed a major work for the mechanical piano. But his enthusiasm was strongly driven by rhythmic and high-tempo possibilities of this instrument that overbids even the most virtuous human pianists. Nancarrow´s history illustrates how the potentials of mechanical music instruments can shape aesthetic ideas and realizations of complex musical scores in a fruitful way. On the other hand, with respect to Beethoven and his limited claviatures, it can also be illustrated how technical limits may not only produce second best solutions but also creative and unexpected improvement of musical ideas. The interrelations between composers and musical instruments are diverse and reach from pragmatic job reasons over new potentials up to unexpected upgrades. This is also true for the current and future use of musical robots. Cultural challenges of musical robots are not only related to composers but to many ways of musical interactions like musicians that are improvising with robots or pedagogues that use robots as another tool for musical education. Musical robots will not replace humans but they can become a part of new musical lifestyles, youth cultures or aesthetic meanings. To enlighten this field of possibilities, gesture oriented studies, which investigate the relations between meaningful body movements and sound, has become the focus of current interdisciplinary research. Aesthetics is not only a matter of musical scores or abstract ideas, but more often embedded in and related to technical practice. If musical robots become part of these practices, they may shape our gestural movements, perceptions and meanings as well as the musical aesthetics and musical lifestyle of composers, musicians, educators or students.

References

Anderson, E. (ed.) 1938: The Letters of Mozart & His Family. Chronologically Arranged, Translated and Edited with an Introduction, Notes and Indices. 3rd Vol., London.

Bevilacqua, F., N. Schnell, N. Rasamimanana, B. Zamborlin & F. Guédy 2011: "Online Gesture Analysis and Control of Audio Processing." In: Solis, J. & K. Ng (eds.) 2011: Musical Robots and Interactive Multimodal Systems, Berlin & Heidelberg, pp. 127-142.

Bruckmann, F. 1993: Zusammenhänge zwischen Klavierbau und Klavierkomposition im Schaffen Beethovens. 2., verb. Auflage, Köln-Rheinkassel.

Buchner, A. 1992: Mechanische Musikinstrumente, Hanau/Main.

Camurri, A. & G. Volpe 2011: "Multimodal Analysis of Expressive Gesture in Musical Performance." In: Solis, J. & K. Ng (eds.) 2011: Musical Robots and Interactive Multimodal Systems, Berlin & Heidelberg, pp. 47-66.

Dittrich, M.-A. 2005: "Die Klaviermusik." In: Leopold, S. (ed.) 2005: Mozart Handbuch, Kassel a.o., pp. 481-559.

Funk, M. & M. Coeckelbergh 2013: "Is Gesture Knowledge? A Philosophical Approach to the Epistemology of Musical Gestures." In: De Preester, H. (ed.) 2013: Moving Imagination – Explorations of Gesture and Inner Movement in the Arts, Amsterdam & Philadelphia, pp. 113-131.

Fürst-Heidtmann, M. 2004: "Conlon Nancarrow." In: Finscher, L. (ed.) 2004: Die Musik in Geschichte und Gegenwart. Zweite, neubearbeitete Ausgabe. Personenteil. Bd. 12, Kassel a.o., col. 897-901.

Godøy, R. I. 2011: "Sound-Action-Chunks in Music." In: Solis, J. & K. Ng (eds.) 2011: Musical Robots and Interactive Multimodal Systems, Berlin & Heidelberg, pp. 13-26.

Godøy, R. I. & M. Leman 2010: "Editor's Preface." In: Godøy, R. I. & M. Leman (eds.) 2010: Musical Gestures. Sound, Movement, and Meaning, New York & Oxon, pp. ix-xi.

Gritten, A. & E. King 2011: "Introduction." In: Gritten, A. & E. King (eds.) 2011: New Perspectives on Music and Gestures, Farnham & Burlington, pp. 1-9.

Herzfeld, G. 2007: "Nancarrows erhabene Zeitspiele." In: Archiv für Musikwissenschaft 64, 2007, No. 4, pp. 285-305.

Hocker, J. 1996: "Mechanische Musikinstrumente." In: Finscher, L. (ed.) 1996: Die Musik in Geschichte und Gegenwart, zweite, neubearbeitete Ausgabe, Sachteil Bd. 5, Kassel a.o., col. 1710-1742.

Hocker, J. 2002: Begegnungen mit Conlon Nancarrow, Mainz.

Ihde, D. 2007: Listening and Voice. Phenomenologies of Sound. Second Edition, Albany.

Irrgang, B. 2007: Gehirn und leibliche Geist. Phänomenologisch-hermeneutische Philosophie des Geistes, Stuttgart.

Irrgang, B. 2009: Der Leib des Menschen. Grundriss einer phänomenologisch-hermeneutischen Anthropologie, Stuttgart.

Ishino, M. & G. Stam 2011: "Introduction." In: Stam, G. & M. Ishino (eds.) 2011: Integrating Gestures. The Interdisciplinary Nature of Gesture, Amsterdam & Philadelphia, pp. 3-13.

Jensenius, A. R., M. M. Wanderley, R. I. Godøy & M. Leman 2010: "Musical Gestures. Concepts and Methods in Research." In: Godøy, R. I. & M. Leman (eds.) 2011: Musical Gestures. Sound, Movement, and Meaning, New York & Oxon, pp. 12-35.

Leman, M. & R. I. Godøy 2010: "Why Study Musical Gestures?" In: Godøy, R. I. & M. Leman (eds.) 2010: Musical Gestures. Sound, Movement, and Meaning, New York & Oxon, pp. 3-11.

Leman, M. 2008: Embodied Music Cognition and Mediating Technology, Cambridge & London.

Ng, K. 2011: "Interactive Multimedia for Technology-Enhanced Learning with Multimodal Feedback." In: Solis, J. & K. Ng (eds.) 2011: Musical Robots and Interactive Multimodal Systems, Berlin & Heidelberg, pp. 105-126.

Noë, A. 2004: Action in Perception, Cambrigde.

Ord-Hume, A. W. J. G. 2001: "Mechanical instrument." In: Sadie, S. & J. Tyrrell (eds.) 2001: The New Grove Dictionary of Music and Musicians. Vol. 16. Second edition, London, pp. 208-213.

Plath, W. 1982: "Vorwort." In: Plath, W. (ed.) 1982: Wolfgang Amadeus Mozart. Neue Ausgabe sämtlicher Werke. Bd. IX/27/2, Kassel a.o., pp. VIII-XXXII.

Solis, J. & K. Ng (eds.) 2011: Musical Robots and Interactive Multimodal Systems, Berlin & Heidelberg.

Solis, J. & K. Ng 2011: "Musical Robots and Interactive Multimodal Systems: An Introduction." In: Solis, J. & K. Ng (eds.) 2011: Musical Robots and Interactive Multimodal Systems, Berlin & Heidelberg, pp. 1-12.

Varela, F. J., E. Thompson & E. Rosch 1997: The Embodied Mind. Cognitive Science and Human Experience. Sixth printing, Cambridge & London.

Essays

Android Robots between Service and the Apocalypse of the Human Being

Gerd Grübler

- It is said that man is the creation of God.

- So much the worse. God had no grasp of modern technology.

Karel Čapek: "R.U.R."

Introduction: Android Robots in Ethical and Social Discussion

Currently, "android" robots sometimes become an issue in ethical and psycho-social investigation. This is in connection with the aim of using robots instead of human beings as "workers" in several fields of service traditionally featured by human-to-human interaction and the question as to whether human-shaped robots would be more easily accepted by the population than abstract-looking and purely functional-shaped robots.[1] Usually, having human-shaped robots has no technical advantage, but in the case that robots are made to directly substitute human agents or to work in environments designed for humans, the humanoid shape might be an appropriate way of minimizing trouble. It might also turn out that because of the general human habit of anthropomorphizing targets of communication, robots with an "android" appearance are the ideal interfaces for human-machine interaction – as Ishiguro (2006) pointed out. For him, it is "Android Science" that deals with the exploration and realization of that type of interface. A "Total Turing Test" – a Turing test that is not confined to intelligent communication but extended to the appearance of a robot – might measure the success in this new field of research and development. There are already some interesting findings. It was shown that people´s behavior in front of an "android" is in many ways similar with their behavior in a human-human communication situation. So, for instance, although knowing that the "partner" was an "android" robot, people could not look into the robot´s eyes when telling a lie; and they reacted to a robot´s smile (Ishiguro 2006, pp. 6-7). Obviously, there must be unconscious reaction patterns that are activated even by "androids," independent of our conscious knowledge that they are not "really" human. Therefore, authentic-looking "androids" might have a good chance of being accepted and respected as partners in communication situations, rather than abstract-shaped robots, and might thus be more efficient interfaces.

However, this very goal-oriented strategy of systematic sophisticated deception might also arouse some criticism: Do we really want to have machines that are able to exploit unconscious human reactions – reactions we do not show voluntarily? Why should we teach robots to capitalize on our irrational suggesti-

1 For some current developments see Menzel & D´Aluisio 2000; Coradeschi et. al. 2006.

bility? It might be such rather skeptical qualms that stand behind the widespread reservation regarding "android" robots that one can find when Western populations are asked to express their respective attitudes. E.g., from a large sample of Swiss citizens being in favor in principle of using robots, only a minority wished that robots have a humanlike appearance (Arras & Cerqui 2005). Broadbent et. al. (2011) showed that in subjects using a concept of "robot" that implies human-shape appearance, blood pressure rises when confronted with a medical robot. The authors suggested that this concept of "robot" makes people more skeptical about and even afraid of robots compared to a neutral concept. So, whether the current interest in humanoid robots is a passing phase or the initialization of a future standard remains an open question (cf. Brooks 2002).

Humanoid robots are, nevertheless, also a good model in machine ethics. Of course, considerations of the rules and abilities one might implement in intelligent devices are relevant in many fields of automatization. But some moral aspects of man-machine interaction are especially well illustrated by human-like machines and these might shed light on several typical fears and confusions. The actual aim of making robots is obviously to substitute and extend human work power, insofar robots are made to be "slaves" without having the cognitive and emotional ability to be slaves: there is no *real* slavery in using robots and intelligent machines. This pattern of thinking of robots as quasi-slaves is certainly a pervading idea. We know that quasi-slaves are today still dull and stupid. It takes too much time to instruct them; they need help too often. So humans long for better "slaves." But at the same time, humans are starting to become anxious and suspicious; they feel that there is a contradiction in this course: If machines become more and more intelligent they will then change their status and become partners – if not masters. Besides rather chauvinistic motives and the fear of being belittled by future machines, another problem comes to light: if it should be possible to construct conscious machines of similar or higher intelligence and sensibility compared to average humans, "slavery" becomes slavery proper. This, of course, implies that intelligence is taken not as sheer computing power but as the ability to possess a comprehensive understanding, which might not be realized without being conscious and having emotions. We might quickly say that this is not at all a realistic option and that such concerns are invalid. However, one should not forget that it is one of the consequences of the broadly shared scientific commonplace that also we humans are only made of matter and that our intelligence is just a function of that matter. So anybody who takes that assumption for granted may not without contradiction plainly refute these concerns.

My central aim here is not to delve deeper into machine ethics, but to briefly document and discuss the, albeit ambivalent, fascination of Western cultures with "android" robots. The following is a rather speculative attempt to comment on this fascination and ambivalence we feel regarding humanoid machines. One has to be aware of the fact that attitudes towards robots might be culturally very different and this might be due to different traditions in historic and moral reasoning (Kaplan 2004; Kitano 2006; Nomura et. al. 2007). Therefore, my interpretations and speculations depend on a certain cultural background and might lose their strength when put into another cultural environment. Though the following is totally open for universal application the author is aware of the Western background the ideas are derived from.

Android Robots in History: Fascination and Fear

Constructing "android" machines has a long tradition (cf. Swoboda 1967; Heckmann 1982; Drux 1988; Richter 1989; Schaffer 1999; Riskin 2003; Sanchez et. al. 2007). Already in antiquity the topos of creating artificial humans was known. Only one example is the story of Pygmalion, the Cyprian king and sculptor who fell in love with one of his ivory statues that, with a goddess´ help, became a real and living woman (Ovid: Metamorphoses X, verse 243 ff.).

From the Middle Ages and the Renaissance we have several relations of the construction of robots that can be found in contemporary fiction as well as in historical accounts by famous people. Unfortunately, all these stories have more the character of a fairy tale or legend and none of the mentioned devices have survived. Talking humanoid heads are a common issue here. Pope Silvester II († 1003) and Roger Bacon (1214-1294) are named among the creators of gadgets of that type. Albertus Magnus († 1280) is said to have constructed a mechanical servant able to act autonomously as a kind of porter. Rather indubitable is that Leonardo da Vinci (1452-1519) designed a humanoid robot in the form of a knight wearing iron armor.

The sheer hype of fascination with "android" robots started in the 18th-century and, in a certain sense, still continues. Here we can find a huge number of artisans and craftsmen who constructed human-shaped automata capable of showing very different instances of human behavior: A monk able to walk around on a table; musicians playing flutes, zithers, drums, flutes, pianos and small organs; writers and drawers using pen, ink and paper to prepare several lines of text or a fine piece of art. Very popular automata were those made by Jacques Vaucancon (1709-1782). Besides the famous mechanical duck, in 1738 he made a human-sized flute player that was able to play twelve different melodies. Vaucancon presented his pieces all over Europe and his exhibitions were extremely popular. The technically most sophisticated and most impressive "android" automata are probably those made by Pierre Jaquet-Droz (1721-1790), his son Henri-Louis (1752-1791), and his apprentice Jean-Frédéric Leschot (1746-1824). Among their pieces is the famous writer, presented in 1775 in Paris. This robot dips a pen into ink and writes a text of up to forty letters. The "Jaquet-Droz" automata have survived, but most of the machines of that golden era of mechanical "androids" have been lost.

In literature, the issue of artificial human beings can often be found (cf. Swoboda 1967; Drux 1988; Sauer 1983; Völker 1994); and the same holds true for cinema. There seems to be two classical ways of showing "androids" here. The first type of artificial man is mere toys or tools of their creators. Of course, sometimes people tend to see more in these machines than sheer mechanisms; horror, confusion, madness and even passionate love might be the results. So is the case in E.T.A. Hoffman´s "Sandman" (1816/17) (Hofmann 1972). The story is of an Italian physics professor and expert in the construction of automata who creates a perfectly designed "android" to impersonate his daughter Olimpia. This robot sings, plays the piano, dances, and says a few words. While most of the people coming into contact with Olimpia react in a rather disgusted manner because of her staring gaze and limited conversation (mostly she is taken to be mentally ill), one overly sensible student at the borderline of schizophrenia falls in love with her and spends a lot of time feeding his fantasies by talking to the puppet. Finally, when his beloved Olimpia´s artificial character is discovered, the student´s madness breaks out and he commits suicide after having attacked and nearly killed his real (human) bride. Motives from Hoffmann´s "Sandman" have also found their place among the standard pieces of ballet and opera: In Léo Delibes´ ballet "Coppélia" (1870) as well as in act one of Jacques Offenbach´s "The Tales of Hoffmann" (1881) the story of Olympia is taken up.

In Hoffmann´s "Automatons" (1819/21) the existence of an "android" Turk answering visitor´s questions – the description is shaped after the famous chess player[2] – is made an occasion to condemn the attempt to build mechanical humans in general. Especially one of the protagonists experiences subtle horrors:

> "All figures of that description, said Lewis, which can scarcely be said to counterfeit humanity so much as to travesty it – mere images of living death or inanimate life are in the highest degree hateful to me. […] it is the oppressive sense of being in the presence of something unnatural and gruesome; and what I most of all detest is anything in the shape of imitation of the motions of Human Beings by machinery. I feel sure this wonderful, ingenious Turk will haunt me with his rolling eyes, his turning head, and his waving arm, like some necromantic goblin, when I lie awake of nights […]." (Hofffmann 1908, pp. 355-356).

2 The mechanical chess-player, dressed in Turk clothes, was made by Wolfgang von Kempelen and first presented to an audience containing among others Maria Theresa empress of Austria-Hungary in 1770. Kempelen presented his apparatus all over Europe. Later on the machine was purchased by Johann Nepomuk Maelzel and presented in Europe and America till 1838. Though this automaton was only a trick automaton in need of a human performer inside this fact was not definitely known to the public throughout the whole show life of the Turk. So the chess-player was a popular issue of speculation both on possible deception strategies and on the overall possibility of performing mental achievements by mechanical means (cf. Standage 2002).

As he is a musician his attitude towards the manifold contemporary "androids" playing instruments[3] is particularly elaborated:

> "The attempts of mechanicians to imitate, with more or less approximation to accuracy, the human organs in the production of musical sounds, or to substitute mechanical appliances for those organs, I consider tantamount to a declaration of war against the spiritual element in music; but the greater the forces they array against it, the more victorious it is. For this very reason, the more perfect that this sort of machinery is, the more I disapprove of it; and I infinitely prefer the commonest barrel-organ, in which the mechanism attempts nothing but to be mechanical, to Vaucauson´s flute player, or the harmonica girl." (Hoffmann 1908, pp. 373-374)

However, within this first paradigm artificial humans convey the attitude and purpose their creators had in mind; they do not really have a "life of their own" – even if it might sometimes seem so. So it is also in Auguste Villiers de l´Isle-Adam´s novel "Tomorrow´s Eve" (1886). Here Thomas A. Edison creates an "electro-human being" (Villiers de l´Isle-Adam 2001, p. 123) as a favor to his young friend, the rich Lord Ewald. This gentleman had fallen in love with a young woman of exceptional, classical beauty. Unfortunately, her "soul" or character is so mediocre and vulgar that Ewald is in a desperate state of mind when he comes to Edison´s home and is ready to commit suicide. The inventor proposes copying the beautiful woman as an "android" and to implement gestures, conversations and habits into this machine that would not contradict her beauty. Though Ewald is skeptical about the possibility of having a "real" relationship with an unconscious puppet Edison manages to convince him that love and especially long-lasting love consists only of the perpetuation of illusions about partners, not in the contact with them as they "really" are. Though there is no happy end in this story (the "android" is lost in a shipwreck only few days after her completion) the project as such succeeds, the illusion is perfect and Ewald is completely satisfied.

In Fritz Lang´s classical movie "Metropolis" (1927) an "android" robot in the shape of a woman is sent out to spoil the organized opposition of the heavily exploited workers in the "Metropolis" state – a scenario obviously based on the "Myth of *Pandora*" (Hesiod: Theogony, 560-612; Works and Days, 60-105), the artificial woman made by *Hephaistos* and sent out to spoil mankind. We find the same in James Cameron´s movie "The Terminator" (1984). At the very beginning of part one, the viewer is taught that machines are neither good nor bad – they just function and execute what they are programmed to do. Thus, the same type of machine, "T-800 Model 101," programmed to kill the central figures of the movie in part one is programmed to protect them in parts two and three.

---

3 Even Beethoven composed music for an automaton orchestra, designed by Maelzel, the second owner of the chess-player (Standage 2002, pp.113-119). See also the chapter "Mozart to Robot – Cultural Challenges of Musical Instruments" (Michael Funk & Jörg Jewanski) in this book.

On the other hand, we can find a second typical way of dealing with humanoid machines and other artificial humans in literature and cinema. Here the machine becomes a really autonomous agent having intentions, interests and feelings itself and therefore acts in a way not foreseen by its creators and often against their interests. We can already find this in Mary Shelley's "Frankenstein" (1818), although here no "android" in the proper sense is concerned. About one hundred years later, we find it again in Karl Hans Strobl's short story "The automaton of Horneck" (1904), in the short story "Moxon's Master" (1909) by Ambrose Bierce and in Karel Čapek's play "R.U.R." (1920) (Strobl 1923; Bierce 1984; Čapek 2004). We are told by Strobl of an extravagant gentleman from the times of romanticism. He not only restores his old castle to the full authenticity of the Middle Ages but also has the desire to hear a prisoner in the dungeon moaning and rattling with his chains during the night. Due to a lack of volunteers he engages an expert in mechanical automata to build an "android" robot to take on the role of prisoner. The automaton, then, is chained and sometimes even flogged to the gentlemen's delight. However, at the end the automaton escapes and, after having shackled his owner in the dungeon, is later found in the gentleman's bed. Bierce describes the tragic death of a technician who was brutally choked by his "android" chess automaton after he had defeated the machine in the game. In Čapek's play the organically manufactured robots (the term "robot" originates from this text) finally engage themselves in a revolution against mankind and kill all human beings. Ray Bradburry's short story "Marionettes, Inc." (1949) (Bradbury 1967) relates similar troubles: This great factory produces "android" robots as copies of existing people who might use them as proxies in life. One of the protagonists just after having decided to purchase one must observe that he already has one – his loving wife. Even worse, the other protagonist's robot is no longer ready to spend its time when not in use in a dark cellar box. Instead it locks his owner in the box and travels to Rio with his wife, whom it has fallen in love with. This schema is taken up again in Ridley Scott's movie "Blade Runner" (1982) in which so-called replicants – "androids" used as slaves in space missions – return to Earth in a violent manner in order to find a way to extend their purposely very short life time of only four years. In "I, Robot" (2004), the movie by Alex Proyas based on short stories by Isaac Asimov (Asimov 1950), the central computer of a big factory producing household robots manages to organize humanoid robots in such a way as to establish a kind of eco-dictatorship over mankind. But there is one more sophisticated robot, called Sonny in the movie, which is able to understand why people prefer not to live under a dictatorship and therefore helps the human protagonists of the story to destroy the supercomputer behind the robot attacks. Some possible moral consequences of creating artificial consciousness are illustrated in Steven Spielberg's movie "A.I. – Artificial Intelligence" (2001). Here it is rather the robot that is the victim of unethical human conduct. The movie shows the fate of an "android" robot in the shape of the little boy David that was made just to "love his mom." When he is abandoned by his "mom" in the forest, for him a never-ending psychological torture sets in. In parallel with Carlo Collodi´s "Pinocchio" (1881) he roams an apocalyptic world trying to find a way to become a "real boy" and by that regain his mom´s love. More lucky is the robot in Chris Columbus´s "Bicentennial Man" (1999), a movie again based on a robot story by Asimov (Asimov 1976). Here the household robot Andrew, model "NDR-114," due to some technological aberrations develops besides his love for Italian opera several features of creativity and autonomous thinking. Supported and humanistic-educated by his owner he manages to lead his own life, earn his own money, and become independent. But then, having seen several generations of "his" family come and go, Andrew decides that he would rather die as a human than live forever as a machine. He spends his money on research into the artificial breeding of biological organs and finds a way not only to substitute human organs with similar ones but also to substitute the mechanical structure of his body with biological components. By doing so, he finally becomes a mortal being and is accepted as human when he dies. The not-so-successful, though for a while promising, reverse strategy is shown in Frank Oz´ movie "The Stepford Wives" (2004) in which women are transformed into servile cyborgs. Here it is one husband´s loving conscience that stops this process so that the remote-controlled wives can get their free will back.

In sum, this brief, and not nearly exhaustive, overview shows that "android" machines have been an issue in Western culture right from the beginning and have constantly and increasingly been in the focus of the public at least since baroque times. Nevertheless, the feelings and attitudes towards humanoid machines featured in literature and cinema are ambivalent. "Android" robots are depicted as sophisticated toys, powerful tools and potentially good fellows, but also as demonic threats to and the fiercest enemies of mankind.

Man – Machine

Obviously, in the early modern age "android" robots could have been taken as mere toys and exhibition pieces displaying the technical abilities of their makers. But at least in the history of Western thinking authors have always felt that "android" machines also allude to rather metaphysical dimensions.[4] Here we meet the voluminous corpus of philosophical reflections on the topic that man himself is a machine. Instead of seeing humanoid machines as artificial *imitations* of human capacities, in these texts they are seen as, albeit still-imperfect, *simula-*

4 Of course, this statement does not claim for totality and focuses rather on philosophical and scientific authors. As Voskuhl has shown, in the late 18th-century there was also a mass production of trivial texts on "android" automata completely devoid of metaphysical or anthropological implications (Voskuhl 2007).

tions or models of their human makers and of beings essentially cognate with them.

Scientists like Andreas Vesalius (1514-1564), William Harvey (1578-1657), and Giovanni Alfonso Borelli (1608-1679), explained the human body explicitly in terms of contemporary technology, meaning in mechanical and pneumatic paradigms (Vesalius 1543; Harvey 1628; Borelli 1680). In René Descartes′ writings we find the philosophical echo of these ideas. For him, human bodies are machines though the essence of human beings, the immaterial soul, is not touched by this fact.

> "Nor will this [that bodies are machines – G.G.] appear at all strange to those who are acquainted with the variety of movements performed by the different *automata*, or moving machines fabricated by human industry, and that with help of but few pieces compared with the great multitude of bones, muscles, nerves, arteries, veins, and other parts that are found in the body of each animal. Such persons will look upon this body as a machine made by the hands of God, which is incomparably better arranged, and adequate to movements more admirable than is any machine of human invention." (Descartes 2010, p. 58)

And this means that

> "the body of a living man differs as much from that of a dead one, as a watch or any other AUTOMA (that is any kind of machine that moves of itself) wound up, having in itself the corporeal principle of those motions for which it was instituted, with all things requisite for its action, and the same watch or other engine when it is broken and the principle of its motion ceases to act." (Descartes 1659, pp. 4-5)

Probably the most well-known text along these lines is Julien Offray de La Mettrie′s (1709-1751) "L′Homme Machine" of 1749. While Descartes took the soul to be a separately created immaterial essence connected to the body, for La Mettrie the soul is not a separate entity but mind, mood and morals directly emerge from the material parts of the body.

> "But since all the soul′s faculties depend so much on the specific organization of the brain and of the whole body that they are clearly nothing but that very organization, the machine is perfectly explained!" (La Mettrie 1996, p. 26.)

Thus, while for Descartes the human body is a machine, for La Mettrie the whole human being is a machine. It is worth noting that in reading Descartes as well as in La Mettrie we find explicit references to contemporary craftsmanship aiming at the construction of "android" automata.[5]

There is, on the other hand, a tradition of hostile reactions to this type of thinking. If man is a machine and if man is a mere material being then there is actually no space for the soul, morality or religion. Thus, arguing against the materialism of the man-machine hypothesis has been the domain of conserva-

---

5 For Descartes cf. the statements above, for La Mettrie cf. (1996, p. 34) where he refers to Vaucancon′s flute-player.

tism and romanticism, and so it has been for a long time. It seemed that if man is a machine then all hope for salvation had to be abandoned. However, my impression is, on the contrary, that the religious attitudes and the strong hope for salvation are exactly on the side of the proponents of the man-machine hypothesis and that this becomes explicitly obvious in the writings of several current A.I. and robotics researchers. This is even more than saying, as Sloterdijk (2001, p. 357) did, that the satisfaction coming from the ability to create machines is bigger than the mortification machines are to humans. What I wish to stress is that it seems that the hope coming from the ability to create machines, together with the assumption that humans themselves *are* machines, is not only bigger than any narcissistic mortification, but is today stronger than all hope deriving from other sources. I do not claim that this hope is well justified; but I claim that it is the drive behind the current movement of hyper-technicalization in general and behind the fascination with human machines in particular.

Already in the 17th-century when we can find the idea that God´s creation works like a very sophisticated clockwork, this idea did not lead to any opposition against the Christian religion, but rather led to religious reformation. Now God as the artisan of this big clockwork realm could be admired even more, the more scientists went into the depth of human and animal bodies and showed the complexity of those "mechanisms." In Henry Power´s "Experimental Philosophy" of 1664 we find the sequence:

> "These are the days that must lay a new Foundation of a more magnificent Philosophy, never to be overthrown: that will Empirically and Sensibly canvass the Phaenomena of Nature, deducing the Causes of things from such Originals in Nature, as we observe are producible by Art, and the infallible demonstration of Mechanicks: and certainly, this is the way, and no other, to build a true and permanent Philosophy: For Art, being the Imitation of Nature (or, Nature at Second Hand) it is but a sensible expression of Effects, dependent on the same (though more remote) Causes and therefore the works of the one, must prove the most reasonable discoveries of the other. And to speak yet more close to the point, I think it is no Rhetorication to say, That all things are Artificial; for Nature itself is nothing else but the Art of God. Then, certainly, to find the various turnings, and mysterious process of this divine Art, in the management of this great Machine of the World, must needs be the proper Office of only the Experimental and Mechanical Philosopher." (Power 1664, pp. 192-193)

This means that we can, through science, understand step by step the way God built the different creatures in the world. And we do so by recreating them ourselves. It is this "knowing-by-reverse-engineering" paradigm which is so pervasive in Western thinking. Now we understand the meaning of the first "android" automata: if we can construct beings able to walk around, to play instruments, to write and so on, we have at least understood something of the way humans are *made*. And this meant to the contemporaries that Western culture – i.e. Western science and technology – is just on the right track.

Today the man-machine hypothesis is common sense and starting point in all life sciences. For most researchers it is enough to accept this assumption as heuristic. However, there is a significant number of authors who have taken up the man-machine hypothesis in its literal meaning. For them, that man is a machine is the precondition for their trans-human perspectives of salvation. Among them we find authors like Hans Moravec, Ray Kurzweil, and Frank Tipler (Moravec 1986; Kurzweil 1999; Tipler 1994). They take the actual human essence as software running on a computing machine. Substituting the current vulnerable bio-computer, the human body inclusive of the brain, by a more durable apparatus for these authors promises the perspective of potential immortality. So Tipler´s theory

> "requires us to regard a 'person' as a particular (very complicated) type of computer program: the human 'soul' is nothing but a specific program being run on a computing machine called the brain." (Tipler 1994, p. 1)

That human mind emerges from a machine in the literal sense, a machine that can be completely known in its functioning and can thus be re-constructed and functionally substituted by human engineers is a basic doctrine here. Alternatively, for Rodney Brooks the future development will be "a merger of ourselves and our robots" (Brooks 2008, p. 75), rather than a sudden replacement of the human being.[6] He expects the process to take much more time than the aforementioned transhumanists say; but in principle he is with them: "I am a machine. So are you." (Brooks 2008, p. 71)

The Apocalypse

Today we do not know whether these ideas are, at least in their overall tendency, realistic and whether the contemporary attempts will pay off in the end. But it seems to me that it is the prime intellectual challenge of today to reach a position on this matter. It is a kind of existential decision whether we believe in technological salvation or not. And we should discuss explicitly this probably strange-sounding question. Not in the sense of talking about all the possible empirical aspects – this is indeed the business of engineers. But we should know whether we assess the probability as high enough to take certain actions. We should know if we can believe in trans-human perspectives and if they attract us. That means that we still have to come to a decision without having the empirical solution on the table, because a future of seeking empirical knowledge and making the crucial collective experiences derives from a decision to do so – and we should at least try to make this explicit. If we really have faith we will not reduce technological activity in order to secure the environment or alternative forms of human life. One might regret this, but compared with salvation nothing really matters. And in fact, this is the tendency of what actually happens. If, on

6 For further information see also Brooks 2003.

the other hand, we decided not to believe in technological salvation, we might question the way things are currently going on in a more radical manner. Not in the sense of becoming techno-phobic; but in the sense of becoming technology electors. As long as we believe in salvation coming from technology we develop and use technology in a fatalistic way: we know that any developmental step might help and that opposition to mainstream use of technology endangers the technological system we have set our hopes on. On the other hand, hopelessness would be the price we had to pay for autonomous and elective use of technology that shuns neither giving up branches of technology nor integrating technology selectively into one´s life and, by that, spoiling the gestalt of the hyper-modern interlinked global system.

This is not the place to discuss these alternatives at length. I think that there are serious philosophical counter-arguments against technological self-salvation. The crucial question "can we organize matter in a way that the essential core of human life emerges?" is a mixed one containing both, empirical and conceptual aspects. Anybody who has children knows that we can organize matter in this way. But technical salvation is, unfortunately, not that easy. A convincing account of technical self-salvation would have to define a core idea of human existence; show the way this existence can be brought into being by human *technological* activities; and explain how the "human core" of existing human beings can be transferred to those entities or preserved by fusion with them. The central key to showing all this is again that man *is* a machine. A machine is a mechanism known in its crucial causal procedures and sub-procedures and, therefore, open for technological setup and further manipulation. Now, if our existence as we know it – or at least those parts we would consider the essential ones for human beings – emerges in a deterministic way from the causal interaction of known and manageable material parts, then we could indeed technologically reproduce or preserve "us." Although this idea seems to be very simple, there are too many ifs involved not to be very skeptical. Counter-arguments might start from the impossibility of defining the essence of humans and end with a sound confutation of determinism as a uniting principle.

However, the ubiquitously expanding technology and the enormous and partially unbelievable achievements of modern technology are strong "arguments" and they are of more or less an inescapable force. This temptation is not easy to overcome and even if philosophy could (and I think indeed can) show that the idea of technological self-salvation cannot be propagated as a sound and sure scenario, it can, on the other hand, not finally prove that technological self-evolution will not finally lead to a kind of salvation. So the contemporary has available strong arguments *against*, but maybe equally strong feelings *for* the technological option. His decision would be a kind of "jump" as Kierkegaard described the way existential decisions happen. My impression is that the contemporary ambivalent attitude towards "androids" (that can partially already be found in romanticism) alludes to this, our, situation immediately before the "jump." There is a well-known topic in contemporary "android" science called the "Uncanny Valley" (cf. Ishiguro 2006; Sofge 2010). The assumption behind this imagination-provoking title is that people evaluate robots the better the more they look human-like – up to a certain level. At that point the score of evaluation goes drastically down (this is the valley) to rise again when absolute perfection would be reached. The "Uncanny Valley" is probably a speculative thought experiment rather than a systematically established fact (Sofge 2010).

It meets, however, an interesting point. Certainly, one might explain the "valley" as a sheer psychological effect: At a certain point our perception of a machine looking like a human being turns into the perception of a human being looking like a machine. The first perception would be a rather friendly one: it is only a machine, but it has a lot of likeable human features. While the second perception would imply disappointment: it pretends to be human but it is only a machine. But I think that, after having talked about the importance of the man-machine issue, we can give the "Uncanny Valley" an additional "metaphysical" meaning that would explain why the valley is really *uncanny*. What I mean is the moment of alteration of the status of something that we experience accompanied by metaphysical horror. The central examples would be the sudden and unexpected transition from life to death and, at least hypothetically, vice versa. It is the transition between something and somebody that raises fears and horrors. In many cultures we can find legends about dead people coming back to life to exist as "Zombies," as living dead. Several aspects of traditional funeral ceremonies are explicitly measures to prevent death from leading to such an existence. So there seems to be some anthropological universality in this topic. I think that the "Zombie-metaphor" can be a good model to illustrate our attitude to "androids" and our situation before the jump. What actually is horrible about "Zombies?" The figure of the "Zombie" is the most horrible being imaginable because it both conquers but also denies the world of our regular existence. "Zombies" are physically as present as the living, but the communicative bridge between them and us has been removed. The existence of the "Zombie" cannot be explained; it refuses all rationalization. We cannot translate the "mindless intentionality" of the "Zombie" into the terms of the order we know. In reality, "Zombies" do not exist; they are a borderline phenomenon existing in fiction only. "Zombie-like" cases are declared finally: a person suddenly woke up in pathology was not really dead but victim of false diagnosis and so on. So the "Zombie" has to decide whether he comes back as a living being or he disappears into death. As a "Zombie" he cannot stay.

I think that this has many things in common with the situation of the "android" robot and the man-machine problem. Unlike the usual case, the "android" robot would have the ability to stay at the borderline for a long time, in the "Uncanny Valley," and this is the horror it embodies. The "android" might leave the

"Zombie" state when clearly recognized as a machine (left side of the "Uncanny Valley") or when finally accepted as human or human-equivalent being (right side of the "Uncanny Valley"). Taken the "android" as the simulacrum of the man-machine problem, the "Uncanny Valley" is something that lies between us and salvation. We have to pass through this landscape while transforming the human by technological means. And here the horror emerges: We are approaching the borderline, not knowing whether we can and, if yes, in what direction we can leave the "Zombie-state." We have to risk our entire existence and to face our biggest fears for the chance of the biggest victory. So, an answer to the core question whether man is machine, is no less than the apocalypse of the human being. And approaching, awaiting and finally giving this answer is consequence and final destination of Western culture. The apocalypse is the moment of discovery when a riddle is solved, when suddenly everything becomes clear and obvious. So if we finally knew that man is machine, things would change completely. If we would finally know that this is not true, thinks would change also, even if not completely. And this seems to be our situation at the moment: we are longing for the apocalypse, hoping, doubting, and not being sure what it will be revealing. We are in front of the valley and ready to jump. We feel too close before destiny to surrender. But at the same time we are afraid of losing too much when failing. Here the ambivalence is center stage again – the ambivalence we have found in the attitudes towards humanoid machines. Technology is strong, perhaps, hopefully, strong enough. But what if we trap ourselves in the "Uncanny Valley," and are not able to escape to the "right" side?

Of course, building humanoid robots is not the only and perhaps not the central field in which the apocalypse is currently attempted to trigger. Projects aiming at the simulation of the human brain as well as research on the fusion of human neuronal structures and computers also belong in that category. But humanoid machines are certainly the most telling and fantasy-provoking pieces of technology embodying the man-machine issue which is the only really apocalyptic issue of Western culture.

References

Arras, K. O. & D. Cerqui 2005: "Do We Want to Share Our Lives and Bodies with Robots? A 2000 People Survey." In: Technical Report Nr. 0605-001, Autonomous Systems Lab., Swiss Federal Institute of Technology Lausanne (EPFL), pp. 16-18.

Asimov, I. 1950: I, Robot, New York.

Asimov, I. 1976: The Bicentennial Man and Other Stories, New York.

Bierce, A. 1984: "Moxon´s Master." In: The Complete Short Stories of Ambrose Bierce, Lincoln, pp. 89-97.

Borelli, A. 1680: De Motu Animalium. 2nd Vol., Rome.

Bradbury, R. 1967: "Marionettes, Inc." In: The Illustrated Man; New York a.o., pp. 156-161.

Broadbent, E., Y. I. Lee, R. Q. Stafford, I. H. Kuo & B. A. MacDonald 2011: "Mental Schemas of Robots as More Human-Like Are Associated with Higher Blood Pressure and Negative Emotions in a Human-Robot Interaction." In: International Journal of Social Robotics, 3/3 (2011), pp. 291-297.

Brooks, R. 2002: "Humanoid Robots." In: Communications of the ACM, 45/3, March 2002, pp. 33-38. (Available at: http://delivery.acm.org/ [accessed April 20, 2012])

Brooks, R. 2003: Flesh and Machines: How Robots Will Change Us, New York.

Brooks R. 2008: "I, Rodney Brooks, Am a Robot." In: IEEE Spectrum online: http://spectrum.ieee.org/computing/hardware/i-rodney-brooks-am-a-robot/0 (accessed April 18, 2012)

Čapek, K. 2004: R.U.R. (Rossum´s Universal Robots), New York a.o.

Coradeschi, S., H. Ishiguro, M. Asada, S. C. Shapiro, M. Thielscher, C. Breazeal, M. J. Mataric & H. Ishida 2006: "Human-Inspired Robots." In: IEEE Intelligent Systems, 21/4, July 2006, pp. 74-85.

Descartes, R. 1659: The Passions of the Soul, London.

Descartes, R. 2010: Discourse on the Method of Rightly Conducting the Reason and Searching for Truth in the Sciences, Wildside Press.

Drux, R. (ed.) 1988: Menschen aus Menschenhand. Zur Geschichte der Androiden, Stuttgart.

Harvey, W. 1628: Exercitatio Anatomica de Motu Cordis et Sanguinis in Animalibus, London 1628.

Heckmann, H. 1982: Die andere Schöpfung. Geschichte der frühen Automaten in Wirklichkeit und Dichtung, Frankfurt a. M.

Hoffmann, E. T. A. 1908: "Automatons." In: The Serapion Brethren. Vol. I. (tans. by M. A. Ewing), London, pp. 352-382.

Hoffmann, E. T. A. 1972: "The Sandman." In: Tales of Hoffmann (trans. by L. J. Kent & E. C. Knight), Chicago, pp. 93-125.

Ishiguro, H. 2006: "Interactive Humanoids and Androids as ideal Interfaces for Humans." In: Proceedings of IUI, pp. 2-9. (Available at: http://delivery.acm.org [accessed April 20, 2012])

Kaplan, F. 2004: "Who is afraid of the humanoid? Investigating cultural differences in the acceptance of robots." In: International Journal of Humanoid Robotics 1/3 (2004), pp. 1-16. (Available at: http://www.csl.sony.fr/downloads/papers/2004/kaplan-04e.pdf [accessed April 20, 2012])

Kitano, N. 2006: "'Rini': An Incitement towards the existence of Robots in Japanese Society." In: International Review of Information Ethics (12/2006), pp. 78-83. (Available at: http://www.i-r-i-e.net/inhalt/006/006_Kitano.pdf [accessed April 20, 2012])

Kurzweil, R. 1999: The Age of Spiritual Machines, New York.

La Mettrie, J. O. de 1996: "Machine Man." In: Machine Man and other Writings, Cambridge, pp. 1-40.

Menzel, P. & F. D´Aluisio 2000: Robo sapiens: evolution of a new species, Cambridge (MA).

Moravec, H. 1986: Mind Children. The Future of Robot and Human Intelligence, Pittsburg.

Nomura, T., T. Suzuki, T. Kanda, J. Han, N. Shin, J. Burke & K. Kato 2007: "What People Assume about Robots. Cross-Cultural Analyses between Japan, Korea, and the United States." In: Nilanjan S. (ed.) 2007: Human-Robot Interaction, Vienna, pp. 275-288. (Available at: www.intechopen.com/download/pdf/595 [accessed April 20, 2012])

Power, H. 1664: Experimental Philosophy, London.

Richter, Siegfried 1989: Wunderbares Menschenwerk. Aus der Geschichte der mechanischen Automaten, Leipzig.

Riskin, J. 2003: "The Defecating Duck, or, the Ambiguous Origins of Artificial Life." In: Critical Inquiry 29/4 (Summer 2003), pp. 599-633. (Available at: http://www.stanford.edu/dept/HPST/DefecatingDuck.pdf [accessed April 18, 2012])
Sánchez, M. F. M., R. F. Millán, B. J. Salvador, R. J. Palou, E. F. Rodríguez, F. S. Esquena & M. H. Villavicencio 2007: "History of Robotics. From Archytas of Tarentum to Da Vinci Robot (Part 1)." In: Actas Urol Esp. 2007 Feb, 31(2), pp. 69-76.
Sauer, L. 1983: Marionetten, Maschinen, Automaten. Der künstliche Mensch in der deutschen und englischen Romantik, Bonn.
Schaffer, S. 1999: "Enlightened Automata." In: Clark, W., J. Golinski & S. Schaffer (eds.) 1999: The Sciences in Enlightened Europe, Chicago, pp. 126-165.
Sloterdijk, P. 2001: "Kränkung durch Maschinen." In: Nicht gerettet. Versuche nach Heidegger, Frankfurt a. M., pp. 338-366.
Sofge, E. 2010: "The Truth about Robots and the Uncanny Valley: Analysis." In: Popular Mechanics, January 20, 2010: http://www.popularmechanics.com/technology/engineering/robots/4343054 (accessed April 18, 2012)
Standage, T. 2002: The Mechanical Turk, London.
Strobl, K. H. 1923: "Der Automat von Horneck." In: Eine Auswahl seiner Erzählungen, Wien & Leipzig 1923, pp. 67-86.
Swoboda, H. 1967: Der künstliche Mensch, München.
Tipler, F. 1994: The Physics of Immortality, New York.
Vesalius, A. 1543: De humani corporis fabrica, Basel.
Villiers de l'Isle-Adam, A. 2001: Tomorrow's Eve, Champaign (IL).
Völker, K. (ed.) 1994: Künstliche Menschen. Dichtungen und Dokumente über Golems, Homunculi, lebende Statuen und Androiden, Frankfurt a. M.
Voskuhl, A. 2007: "Producing Objects, Producing Texts. Accounts of Android Automata in Late 18th-century Europe." In: Studies in History and Philosophy of Science 38/2, pp. 422-444.

Joseph Weizenbaum, Responsibility and Humanoid Robots

Kerstin Palatini

Joseph Weizenbaum is often mistaken as computer critical (Weizenbaum & Wendt 2006, p. 7); he has not been arguing against computer at all, but for a humane society, based on rational used computer technical support. Therefore, computers do not have to be humanlike. He is observing related developments within Artificial Intelligence (AI) very critical. And if he is watching the current[1] AI – research, especially humanoid robotics, he questions their sense and profession.

"Not *a little bit* negative. ..."

(Joseph Weizenbaum´s relationship to humanoid robots.)

"No, I am *very* negative." proclaims Joseph Weizenbaum after he got criticized by the claim of the moderator Loïc Le Meur, which intends that Weizenbaum occurs quiet negative towards the current informatics development during his attendance at a panel discussion 2008 to the "World Economics Forum" in Davos, one of his last public appearances.[2]

People invited the famous informatics pioneer to Davos, who had served and taught for years at MIT, and who is perhaps even seen as hero of informatics, especially for Artificial Intelligence. And now this clear critical claim. Hereby, the reaction of Weizenbaum could have been foreseen as far as if you had followed his work, which displayed his increasing negative view on the current informatics development, especially on the AI-research as well. In his latest book, which is structured as interview, Weizenbaum tells us what makes humans to be human. He intends to realize that humans are social beings, who are effected by their experiences, needing mutuality, using languages to communicate and being characterized by their mortality and finite nature (Weizenbaum & Wendt 2006, pp. 103ff.).

For this reason, the essence of humanity implies something totally different for him in contrast to his AI-colleagues Hans Moravec or Marvin Minsky. They claim the essence of humanity would be information, which can be variously perceived and programmed as input into a computer, rigorously taken apart and putted together again, as well as reproduced; like the "human being could be represented by an endless chain of bits" (Weizenbaum & Wendt 2006, pp. 105-106). In consequence it is no longer impossible for the scientists to move for-

1 Joseph Weizenbaum died on March 5th 2008 near Berlin.

2 The discussion was about virtual Worlds, in particular about Second Life, link to the video: http://www.youtube.com/watch?v=E198IynGbg0 (accessed July 25, 2011).

ward to the infinite reproduction of the human being (as well as following into immortality).

Weizenbaum criticizes the faith in fairy tales of people, who fall for the fascination of future promises by these and other distributors of Artificial Intelligence (Weizenbaum & Wendt 2006, p. 128). Further, he exposes the fact of the development of special application systems and so called expert systems, also, while people talk about Artificial Intelligence within most cases. Behind the "computer intelligence" there is a "pure quantity," "raw power" of computers in terms of an enormously increased capacity of processing power and storage performance (Weizenbaum & Wendt 2006, pp. 125-126). They enable e.g., to land planes, playing chess, executing difficult medical operations, installing auto parts, and so on. But they are not capable of creativity, comprehension, generating meanings neither being quiet responsible:

> "Responsibility is namely no technical category, but a social one. And the current status of our society is characterized by denying responsibility. Furthermore: Our society did intentionally develop technology, so responsibility actually became so much divided that no one actually has it." (Weizenbaum & Wendt 2006, p. 32; see also p. 115)

Responsibility is getting replaced by automation. "Any doubt [related to contents and consequences – K. P.] disappears in the context of the fast and goal-orientated acting." (Weizenbaum & Wendt 2006, p. 34) "The fastest solution for tiniest problems" belongs to that, too, which reminds us on the studied formula: "Problem – solution – expert – means, headache – Aspirin. [...] Someone uses immediately aid" (Weizenbaum & Wendt 2006, p. 84), and gives up thinking about the main reasons and long termed consequences of aid. In other words it means passing on the responsibility for our own acting.

But responsibility cannot be passed on to technique, formula and technologies. Quotation by professor Irrgang:

> "Deliberating or judging computer as moral beings would easily open up backdoors for the responsibility dilemma. Further it might enable deporting strategies of the originally moral responsibility. [...] In this case, the computer would be the guilty one." (Irrgang 2010a, p. 253)

Experts in the field of Artificial Intelligence, who try to develop humanoid robots, humanlike beings or artificial life might probably see themselves as awakener of old myths and legends and maybe one of those, who helped the humans to realize the dream of flying into space or diving into depth. All that made possible by engineering. Unlike to that the content of myths and stories about creation of humanlike beings, mostly about nightmares and horror visions, reminding on stories about the "Golem" or "Frankenstein´s" monster (Weizenbaum & Wendt 2006, p. 86). In the light of the unruleable creation even Goethe initiates his "Zauberlehrling" to exclaim: "The ghosts, those I have called, now I do never get rid of them."

The excitement, also for Joseph Weizenbaum is that the "dream exists, and the humans work on it," "specifying the idealistic human being and later producing him." The Artificial Intelligence is increasing the potential to realize this dream. Weizenbaum´s doubt starts here within the term of intelligence, which he claims to be "almost an illusion," while it is linked to AI (Weizenbaum & Wendt 2006, p. 87). Continuing, he describes it as "an insane dream of Artificial Intellgence to create a machine, a robot, which is transforming into a human being" (Weizenbaum & Wendt 2006, p. 104).

But things which sound like Hollywood – horror or science fiction fantasies have already become reality at science laboratories all over the world. Humanoid robots are very popular, especially in Japan (Irrgang 2010a, pp. 250-251). The Japanese scientist Hiroshi Ishiguro gets introduced during the documentary movie Plug and Pray.[3] He has created a robot – alter – ego called "Geminoid" (Ishiguro 2010). Ishiguro appears to be very proud of his "android" robot. So he presents himself somehow advertisingly efficient while he is wearing the same outfit as his image. Besides the remarkable proudness of the creator referring to his work there is something pitiful as well as compassionate within this scene. Especially, if the scientist is saying, that he can well imagine this robot could play with his children once Ishiguro would be on business trip, which happened quiet often.

This reminds us on the baroque mechanical toys of earlier rulers, e.g. the precisely copied birds in their golden cages, such as they can be seen today at the Green Vault in Dresden (Germany), gratefully gifted by the king of Saxony August the Strong. Here, we have to realize that these valuables have not been collected by the collector´s passion during the 18th-century, but more for representation of wealth and power. Their mechanical moving and the copied noises by the pieces of jewellery pleased the entire royal household as well as foreign lords. This has led to amazement and admiration of their owner (August II.), caused by the expensive attractions. In the light of that it appears nothing much has changed since then: the proudness of creators to their artefacts, the vanity of some owner and the meaning as sign for power have endured.

Alongside the effort of the perfect imitation came up the effort to optimize and idealize nature which did not even stop in front of creating human beings. Weizenbaum distinguishes two ways of solving this "Golem-dream:" "We try to produce gods or further trying to become gods. At least we construct and build things, so we can honour those [...]." (Weizenbaum & Wendt 2006, p. 110) The development of such beings is connected to current duties of research, to ques-

3 Since November 11th 2010 at the movie theatres: "At Plug & Pray the regisseur and Grimme Award winner Jens Schanze is opening a dialogue between fanatic researcher and an old wise professor about the question, what is mankind truly about. The movies lately ends with a senseful closing speech for humanity and the reverence in spite of natural mystery of life and death." (http://www.plug-pray.de/ [accessed April 19, 2012])

tions which researchers and scientists do not only ask themselves, but questions which get asked to the researcher by the society and politics. Depending on interests are those questions supported or not. Relating to decide about questions of research Joseph Weizenbaum points out:

> "We [scientists] decide the selection, what is bound to values. She is strongly affected by social circumstances in which we live. Therefore, it is no coincidence that researchers have come to certain problems." (Weizenbaum & Wendt 2006, p. 11)

Only the embedding of technology and technological development within their surrounding context (Irrgang 2008) are exemplarily described here. But at the same time there are important and responsible roles of technology inventors and creators at the production process of technical artefacts included, too (Irrgang 2010b). Because this process is a complex and social process and it comes along with risks. That is why we have to talk about "collective responsibility and accountability" (Irrgang 2008, p. 379) as well as drawing attention to it.

Joseph Weizenbaum´s popular speaking-analyzation-program ELIZA of 1963 (Weizenbaum & Wendt 2006, pp. 89ff.) proved for him what can happen if you pair technical programs with human attributes. There was not even a humanoid shape. Only the ability of the program was enough to be characterized as human conversation by many users. It was able to talk within a context related conversation, meaning "listening" and "respond" and additionally generating those. Weizenbaum recognized here the danger of abusing technology.[4] He had only envisioned displaying information processing by ELIZA.[5] According to that, he has been wondering about (and fearing) the serious use of this program by experts, therapists and human experts at the psychotherapist context[6] of their offices: The belief was to save time while they wanted to "take care of hundreds of patients at the same time" (Weizenbaum & Wendt 2006, p. 90).

That moved Weizenbaum to rethink, because his automized responses were based on scripts and algorithms. He realized by ELIZA that he had overcome a barrier of technology, moving towards humanization. Probably did ELIZA cause for Weizenbaum one of his most remarkable moments of cognition.[7] But beyond that it demonstrated the kind of risk potential what was enclosed within such technological developments.

---

4 After further reports about ELIZA Joseph Weizenbaum published in 1978: "Die Macht der Computer und die Ohnmacht der Vernunft." (The power of computer and the powerlessness of rationality.). Since then Weizenbaum is known as hard critic of Artificial Intelligence (AI) and informatics.

5 "ELIZA" referring to "Pygmalion" by G. B. Shaw or by the flower woman Eliza Dolittle in "My Fair Lady," who keeps on improving her language skills by the instructions of Professor Higgins (Weizenbaum & Wendt 2006, p. 90).

6 A script opportunity by ELIZA, which Weizenbaum had called "Doctor."

7 See Weizenbaums quotation at the end of this essay.

At the end there is a typical anecdote by Weizenbaum: He talked about his experiences during his trip to Japan at the Bauhaus in Dessau in November 2007.[8] He found himself amused by a conversation with a Japanese robotics scientist who had introduced him to a real humanoid caring (service) robot of Japanese future. The issue was just about its long time of production. As Jospeh Weizenbaum said he would likely prefer a dog to keep him company at older age the Japanese researcher responded: a dog-robot can be built in four weeks, it would be totally uncomplicated from a technical point of view. Instead of being capable to interpret Weizenbaum`s response correctly the Japanese scientist obviously appeared to be too much involved into his "omnipotent technological possible world."

Although scientifical authority or fascination of future promises underlies this phenomenon (Weizenbaum & Wendt 2006, p. 128) – often we are victims of modern myths by technology and computer development. We might follow them without rethinking if there would not be some people, who encourage others to reflect their opinions. Of course, those people argue with their own community and politics because they show off threats and risks by promoting their critical view. And this is often the reason for the prejudice to hold back further advancements. Calling on those advancements which appear during AI ambitions to create humanoid robots and making them more humanlike.

What is Philosophy of Technology capable of ...; allowed to do ...; and has to do ... to find *ways out of a programized society*[9]? Firstable, Joseph Weizenbaum should be honoured and we should be grateful for his courage. Because he (as certificated expert on this field) has permanently argued critically with success and experiences at his fields of knowledge and employment; he got himself critical as well as listened. Furthermore, he was fighting for the human part of technology until the end knowing that technology does not have to be artificially intelligent or humanoid designed. Weizenbaum:

> "We all know such moments in which we suddenly realize something what is most essential. We should call it out loud. We should let other people participate at our knowledge." (Weizenbaum & Wendt 2006, p. 67)

Hermeneutics of Technology can "precisely look" (Weizenbaum & Wendt 2006, pp. 57ff.) at decisions, at willing to recognize and realize, at supporting to build mistrust or trust. Philosophers of Technology can methodologically carry on what has been started by Weizenbaum and some of his colleagues.[10] They are capable of something what computers are not: Keep on asking questions, e.g. of

---

8 Weizenbaum spoke infront of the Volume "Informatik und Gesellschaft" (Informatics and Society) on November 10th 2007, after the invitation of the department of informatics at the College of Anhalt at the aula of Bauhaus, Dessau.

9 Referring to the subtitle of the book (Weizenbaum & Wendt 2006).

10 Weizenbaum names some of his appreciated colleagues, such as Noam Chromsky (Weizenbaum & Wendt 2006, pp. 73-74).

values, of responsibility, of current research themes; encouraging public discussions, opening different perspectives, describing structures and deriving contexts and further maybe to realize enlarging and connecting "the islands of rationality." Perhaps those islands will grow up to continents. This would not be possible without knowledge and experience, trust and consciousness of responsibility. Otherwise computers would be capable of it. Referring to humanoid robotics is the following final sentence a cautious, but an urgent one:

> "The life is not much secure. And those things what we are talking about here – the experiment to create artificial life – affects all human beings. They might have big influence." (Weizenbaum & Wendt 2006, p. 88)

References

Irrgang, B. 2008: Philosophie der Technik, Darmstadt.

Irrgang, B. 2010a: Homo Faber. Arbeit, technische Lebensform und menschlicher Leib, Würzburg.

Irrgang, B. 2010b: Von der technischen Konstruktion zum technologischen Design. Philosophische Versuche zur Theorie der Ingenieurpraxis (Technikphilosophie Bd. 22, ed. by Klaus Kornwachs), Münster.

Ishiguro, H. 2010: "The Man Who Made a Copy of Himself." In: IEEE Spectrum, 04/ 2010: http://spectrum.ieee.org/robotics/humanoids/hiroshi-ishiguro-the-man-who-made-a-copy-of-himself/0 (accessed April 19, 2012)

Weizenbaum, J. & G. Wendt 2006: Wo sind sie, die Inseln der Vernunft im Cyberstrom?, Freiburg, Basel & Wien.

Acknowledgement

This essay was inspired by personally meeting Joseph Weizenbaum in 2007, friendly connections with the film-makers of "Weizenbaum – Rebel at Work," Silvia Holzinger and Peter Haas, watching the film "Plug and Pray" directed by Jens Schanze, and, of course, reading Weizenbaum's books:[11]

(Translated by Silvio Wende)

---

11 Poster-Download Plug & Pray: http://www.farbfilm-verleih.de/filme/plug_pray.html. The poster is printed in this Publication "Robotics in Germany and Japan" with friendly authorization of the Entertainment Kombinat GmbH, represented by Jasmin Knich, Berlin (by E-Mail from 21.9.2011). Thanks for kindly support.

„DAFÜR WERDEN WIR ALLE EXKOMMUNIZIERT."
Prädikat besonders wertvoll
PLUG &PRAY
EIN FILM VON JENS SCHANZE

Social Stereotypes as a Guarantee for Proper Human-Robot Interaction? Remarks to an Anthropomorphic Robot Design[1]

Manja Unger-Büttner

Nothing could be more appropriate to introduce the following reasoning than to quote the media philosopher Vilém Flusser – out of a book, whose German title could not be chosen better: "Vom Stand der Dinge" / "The state of things"[2]:

> "The old lever is striking back at us: We have been moving our arms as though they were levers since we have had levers. We simulate that which we have simulated. [...] This striking back on the part of the machines is now becoming clear for all to see: young people dancing like robots, politicians making decisions based on computerized scenarios, scientists thinking digitally and artists using plotters. Consequently, the fact that the lever is striking back will have to be taken into account in the future construction of machines. It is not enough simply to take the economy and ecology into consideration in the construction of machines. We will have to think about the ways in which such machines may strike back at us. A difficult thing to do considering that most machines nowadays are made by 'intelligent machines' and that we ourselves only look from the side-lines, as it were, intervening only occasionally." (Flusser 1999, p. 53)

The design, the surfaces of these new machines may not be completed by other machines in the near future.[3] But automatisms and stereotypes, deep in the thinking of responsible professionals (perhaps totally unconsciously), can be part of the design process. Starting from this parallel, Flusser´s "Looking from the side-lines" will be elaborated for a *closer inspection* for the purpose of this essay,[4] in order to make a small contribution to a problem that Flusser denominated to be one of the design: "What should machines be like if their striking back is not to cause us pain? Or, better still: if it is to do us some good?" Issues like these should be discussed *before* starting to design "stone jackals and supermen." And Flusser asks: "Are designers ready to address them?" (Flusser 1999, p. 53)

---

1 Thanks to Dan Anderson, Auburn/California.

2 The official translation of Flusser´s book sadly is not a direct one: "The Shape of Things: A Philosophy of Design." A more direct translation would be: "The state of affairs. A Philosophy of Design." – but this also does not catch the German connection between "things" and "affairs" (Dinge/Dinge), that Flusser surely was focusing.

3 About the lack of the anthropological dimension, in case that technology would be built only by other technology, see Irrgang 2010, p. 253.

4 Corresponding to the title of the conference and the attendant book "Robotics in Germany and Japan: Cultural and Technical Perspectives," the references for this essay are deliberately taken from the German cultural sphere.

As part of a comprehensive interdisciplinary approach, practiced at the Institute for Philosophy of Technology at the Technical University in Dresden, designers address these issues. Out of their design practice they are trying to find philosophically funded answers. New input for this arose with the workshop "Future of Robotics in Germany and Japan." In this framework, for example, Mark Coeckelberg pointed out, that robots always already have some sort of surface or manifestation, actually some design.[5] Whether this surface appears *human-like* to the beholders, is often a matter of the viewers themselves. The Swiss musician Christian Denisart once said in an interview about the universal tendency of humanizing the artificial environment: "If you add eyes and wheels to a coffee machine, you will bond with it, although it is still a coffee machine." (Ichbiah 2005, p. 48 [Translated by the author, Manja Unger-Büttner.])

Robots certainly do not anthropomorphize themselves: people, rather, tend to relate unfamiliar objects as soon as possible to something familiar. By trying to withdraw everything enigmatic, eerie or disconcerting from objects, people already do the first step to anthropomorphize them. Placing unknown objects into half-knowledge and classifying them as soon as possible with something well-known, is a function of everyday consciousness. Also the daily practice of *reducing* unfamiliar objects into something human-like by implication, illustrates the *ascribing* of human-likeness (Leithäuser 1976, p. 12). Anthropomorphism, in both ways, thus is an act by the viewer.

The appearance of robots is particularly important when designated for the vicinity of humans. Reciprocal relationships between people colloquially are called *social*. The moderation of relationships is one of the key tasks and challenges of design (Höger 2005, p. 158). The social impact of any determination, done by developers of robots, be they engineers or designers,[6] can hardly be overestimated. An adequate design for a proper "personalization" of robots is believed to support and facilitate any sentiment of social bonding to the robots (Becker 2009, pp. 58-59; see also Weber 2009, p. 45). Thus, emotions, mediated through robotic facial expressions and gestures, are a popular topic in current robot research and development. But these expressions *remain* limited to the behavioral level of the robot and have no relation to any "experience level" of it. The aim of these efforts is to make communication with robots more intuitive

5 See also the chapter "Robotic Appearances and Forms of Life. A Phenomenological-Hermeneutical Approach to the Relation between Robotics and Culture" (Mark Coeckelbergh) in this book.

6 The lovely face of the child-like humanoid robot for cognition research, named "iCub," really is a "faithful representation of the infant inspiration." Its design was one of the most important jobs for granting the best human-computer-interaction (see http://www.cicada.manchester.ac.uk/events/workshops/robotics/eveningtalk.pdf [accessed November 6, 2011]). Designer of the face was the Italian Ugo Gallini, from 3D Terza Dimensione (direct information to the author, Manja Unger-Büttner, by email from Giorgio Metta, Italian Institute of Technology, Genoa, July 7, 2011).

and "natural" for the human user and to promote a longer interaction and reliance. To use the term *social robots* for service robots in direct contact with people, can be interpreted as a further step to building more confidence. Certainly, the media scholar Barbara Becker notes a special illusion, which is directly supported by this *term*: notions of a social practice. The assumption, that artificial agents could be equal partners in interaction, can be strengthened this way (Becker 2009, p. 59).

The communication scientists Byron Reeves and Clifford Nass noted that human interactions with computers, television and new media are "fundamentally social and natural." (Reeves & Nass 1996, p. 5). As with interactions in real life, people also expect media to follow certain social and natural rules originally derived from interpersonal interactions and experiences how people interact in and with the real world. It seems as if all these rules could also be applied to media. These interpersonal experiences and traditional social rules are underlying every interaction with media today. Relatively recently, a completely new deviation from the common experience emerged: "Modern media now engage old brains." During the past 200.000 years anything that acted socially, was actually a person, and anything that was observed, was in fact real. Nowadays these things can also happen virtually, communicated through media. Human responses continue to run almost automatically (Reeves & Nass 1996, p. 12).

Many robotics researchers base their theories on Reeves and Nass with their evolutionary reasoning that people *became accustomed* to behaving socially through millennia and now do it towards non-human actors, too – as long as these actors also "interact socially." Therefore, these researchers rate the role of physical and aesthetical appearance of robots as very important (Weber 2009, p. 39). Reeves and Nass emphasize possible impacts of disregarding the rules they found for robot development:

> "A lack of concern for social and natural rules by designers does not mean that the rules will be irrelevant; instead, users will simply be frustrated and unhappy." (Reeves & Nass 1996, p. 255)

Finally, robots with human-like appearance and an option of so-called "mimetic and gestural sensation" are considered to be particularly useful social interfaces. The philosopher, media theorist and technology researcher Jutta Weber emphasizes another reason why humanoid design is considered to be outstandingly effective for robots: because they are rather mobile than other intelligent agents and thus also physically closer to people (Weber 2009, p. 39).

It seems as if Reeves and Nass determined a basis for every "social" robot design and influence many of current designs in the field of robotics with the following sentence: "Stereotypes make a complex world simple, even if they also cause us to miss critical individual differences." (Reeves & Nass 1996, p. 167) – These basic stereotypes are now revealed as problematic by Jutta Weber in her remarkable essay. She notes, how deep these "social" moments and gender stereotypes in robot design are likely exploited to build "better" machines, perceived to be social and intelligent (Weber 2009, p. 44). The small child pattern, for example, is specifically used to trigger a sort of nursing reflex in users. The concept behind this is modelled out of stereotypical assumptions of the traditional mother-infant relationship in the bourgeois nuclear family, in which the woman invested all of her time in household and raising children (Weber 2009, p. 38).

A second stereotype-laden aspect in robot-design is gender. A completely sexless appearance is thought to make users afraid – hence a sexually coded design is quite common. The "android" "Valerie,"[7] originally conceived as a household robot, was designed as "feminine," in order that people would be less frightened by the life-sized robot. Finally, "Valerie" is particularly striking for its gender-specific, almost sexist design, and soon it advanced from a cleaning robot to a mannequin. Jutta Weber remarks a certain constraint behind this seemingly inescapable gendering: robotics researchers underline that language software is not really available without gender-specific characteristics. So they seem simply to subject themselves to restrictions like this. Other constraints can accidentally promote the aim of increasing the interest of girls and women in robots: While designing the small robot doll "Robota," only *female* 30-inch dolls, needed for building a body for "Robota," were available on the market. Jutta Weber asks why, given the high investments for a project like this, there was no opportunity for a specially designed doll. And one can also ask if there is enough research about voices and gender. Weber criticizes the further implicit promotion of traditional gender stereotypes on the example of "Robota": this robot for children has long hair and is dressed in girlish clothes – and one favourite "activity" of her is to dress well (Weber 2009, pp. 39-40).

Barbara Becker, in the same book like Weber, mentioned that "Robota" also was made for interaction with autistic children, based on her minimal facial expressions and gestures. The manageable signals of a robot doll seem to be making autistic children less afraid than the complex, largely non-verbal communication in human society. The children followed the eyes of the robot girl and imitated simple movements. Becker questioned whether such an approach for therapeutic purposes really is legitimate (Becker 2009, p. 58).

Joining Becker´s comments with Weber´s insights into gender stereotypes in robot design, it begs the question if it is appropriate to confront small people with "impaired social interaction and communication"[8] with such an artificial shrinkage of the social world and communication for therapeutic purposes – as all robots currently demonstrate. Thus, the stereotypical anthropomorphic design of robots, a question of outer appearance, is directly leading to questions of social practice, even to *ethical issues*: Barbara Becker notes that dealing with hu-

7 See: http://www.androidworld.com/prod19.htm (accessed September 12, 2011).

8 This is a definition of autism, in: Matson & Sturmey 2011, p. 3.

man-like artifacts can be unproblematic, as long as the user remains a reflective distance from these artificial fellows. For Becker, children are already overwhelmed with this situation, but this seems to be no real problem, as long as the robots have a certain status like plush toys. Dolls and plush toys always have been attributed to personality and children seem to be implicitly aware of their artificiality. The future will determine if this will change with the increasing of facial expressions and gestures of robots (Becker 2009, p. 59).

Considering both Becker`s and Weber`s studies, an increasing responsibility becomes clear, when using reduced gestures and facial expressions for therapeutic purposes or as "social caretakers." Hence, Becker is calling for detailed empirical studies and theoretical reflections on sophisticated social and psychological effects of robotic applications like these (Becker 2009, p. 59). It seems to be important to call these new forms of communication and the projective attributions on the part of users in question.

If used therapeutically or developed only for entertainment or as household help: anthropomorphic robots cannot reflect any social flexibility except the norm due to social stereotypes. They cannot *play* with gender roles – but only objectify gender differences. Jutta Weber identified the assumption of a rule-based social behaviour between people to be a starting point for these developments in robot design. Thus roboticists often use biological and (vulgar-) psychological concepts of sociality and emotionality, based on a (one-dimensional) functionality to map human-machine relationships (Weber 2009, pp. 44-45). This deserves criticism because of its too-shallow picture of the world. Ernst Cassirer, for example, emphasized that any new tool can form not only the outside world but also human self-consciousness (Cassirer 2002, p. 254). That is why an unthinking, stereotype-based anthropomorphism in robotics design may lead into a spiral of abrasion in the perception of the world as well as in the perception of people and oneself.

Jutta Weber detects that the new research in the field of human-robot interaction could pursue a naive, unpretentious attitude towards machines on the part of the users. This assumption she connects with findings from earlier research on artificial intelligence, which point out that the machines do not adapt to humans, but people adapt themselves to the machine (Weber 2009, pp. 43-44). For example, human "users" dealing with robots or sophisticated machines have hard work every day: secretaries still have to use a rather simple, standardized language while using translation or dictation ambiguities – and equal to this, the user of robots has to limit himself to make the interaction "intelligible." Many of the supposed benefits of computers are products of the efficient work of people and often the shortcomings of the machines are actively compensated by these people (Weber 2009, pp. 44-45). Technical systems will never achieve complete accuracy, and that is why a special capability has now been taken into account: the concept of resilience treats a certain error tolerance and adaptability on the part of technics as well as the user (Palatini & Richter 2011, p. 266). Unthinking design of robots could request questionable adaptation services by the user: for example by cementing questionable stereotypes, using a highly simplified language, consideration of the currently mostly limited power of sight and mobility of "social" robots. This service, done by the user, seems not to equate with resilience. Of course, these weaknesses could be teething troubles of robotics, considered to be corrected by steady advancement in the future. But *then* it could be very interesting to look back in time and compare who and in what degree has converged and adapted to whom.

According to the philosopher of technology Bernhard Irrgang, the essence of technology is not scientific knowledge, but the ability of handling technical artifacts ("Umgehen-Können"). Together with the conception of technical artifacts, handling them is not only a science, but rather an art to Irrgang (2007, p. 35). In a positive, people promoting, intellectual understanding of art, it seems questionable for the development of human skills to adapt oneself to technological artifacts when seeking the performance one expects. This raises the question, if human skills and the *art of handling* technology could curtail development, when the *design* of technology is anchoring in generalised social norms.[9]

Erhard Tietel mentions a "mechanization of mental labor" (Tietel 1995, p. 272), which should not be understood only in the common sense, that operations which actually could be done by the human brain, are done by machines. In addition, this mechanization also means that actual unique thought processes of a human being, permanently reshaped through normalized input, can become normalized and normed – and involuntary could be treatened as equivalent to machines. The thinking could become mechanized. And Weber concludes that the change of paradigms, away from rational-cognitive towards "social" machines, does not lead away from a reductionist technology design (Weber 2009, p. 44). Vilém Flusser´s complaint about the "striking-back of the machine" seems to confirm this idea.

The importance and impact of the appearance of our machines have been discussed now only to a limited extent. In active and close cooperation with reflecting sciences, such as philosophy of technology and ethics, designers could try to unveil the frequently unconscious mechanisms in the development of technology. Supported by scientific reflection and interpretation in a real interdisciplinary approach, designers could exercise a phenomenological-hermeneutic balance out of their unique perspective from practice – for making profound decisions between possible targets of anthropomorphic robot design and contradictory, critical consequences. On one hand this could result in an optimization of *human-robot* interaction, on the other hand, as shown above, it concerns effects on *human* and *interpersonal* affairs...

9 Concerning possible fictional elements in social sciences and the "mathematization of society" (Keller 2009, p. 108).

References

Becker, B. 2009: "Humanoide Roboter, emotionale Konversationsagenten. Anmerkungen zu neuen Konzepten in der Mensch-Maschine-Interaktion." In: Friesinger, G. & K. Harrasser (eds.) 2009: Public Fictions. Wie man Roboter und Menschen erfindet, Innsbruck, pp. 52-64.

Cassirer, E. 2002: Gesammelte Werke. Bd. 12: Philosophie der symbolischen Formen, Teil 2: Das mythische Denken, Hamburg.

Flusser, V. 1999: The Shape of Things. A Philosophy of Design (Translated from German by Anthony Mathews), London.

Höger, H. 2005: Design und Politik. Schriftenreihe querfeldein, Bd. 1, Würzburg.

Ichbiah, D. 2005: Roboter. Geschichte _ Technik _ Entwicklung, München.

Irrgang, B. 2007: Technik als Macht. Versuche über politische Technologie, Hamburg.

Irrgang, B. 2010: Von der technischen Konstruktion zum technologischen Design. Philosophische Versuche zur Theorie der Ingenieurspraxis, Berlin.

Keller, F. 2009: "Von Sternen und Menschen. Mathematische Fiktionen von Gesellschaft." In: Friesinger, G. & Harrasser, K. (eds.) 2009: Public Fictions. Wie man Roboter und Menschen erfindet, Innsbruck,pp. 106-119.

Leithäuser, T. 1976: Formen des Alltagsbewußtseins, Frankfurt a.M.

Matson, J. L. & P. Sturmey 2011: International Handbook of Autism and Pervasive Developmental Disorders, New York.

Palatini, K. & Richter, V. 2011: "Resilienz – 'Joy on Error.' Usability als Chance und Ressource." In: Brau, H., A. Lehmann, K. Petrovic & M. C. Schroeder (eds.) 2011: Usability Professionals 2011, Stuttgart, pp. 266-270. (Available at: http://germanupa.de/fachkonferenz/up-2011-1 [accessed April 18, 2012])

Reeves, B. & C. Nass 1996: The Media Equation. How People Treat Computers, Television, and New Media Like Real People and Places, Cambridge.

Tietel, E. 1995: Das Zwischending. Die Anthropomorphisierung und Personifizierung des Computers, Regensburg.

Weber, J. 2009: "Hilflose Maschinen und treusorgende Pflegepersonen." In: Friesinger, G. & K. Harrasser (eds.) 2009: Public Fictions. Wie man Roboter und Menschen erfindet, Innsbruck, pp. 36-51.

Authors and Contact

Dr. Mark Coeckelbergh

Department of Philosophy + 3TU.Centre for Ethics and Technology
University of Twente
P.O. Box 217
7500 AE Enschede
The Netherlands

Webpage: http://www.utwente.nl/gw/wijsb/organization/coeckelbergh
Email: m.coeckelbergh@utwente.nl

Prof. Dr. Michael Decker

Institute for Technology Assessment and Systems Analysis (ITAS)
Postfach 3640
D-76021 Karlsruhe
Germany

Webpage: http://www.itas.kit.edu/english/staff_decker_michael.php
Email: michael.decker@kit.edu

Michael Funk, M.A.

TU Dresden
Institute of Philosophy
D-01062 Dresden
Germany

Webpage: http://www.funkmichael.com

http://tu-dresden.de/die_tu_dresden/fakultaeten/philosophische_fakultaet/iph/tph/Michael_Funk/index_html

Email: michael.funk@tu-dresden.de

Dr. Dr. Gerd Grübler

TU Dresden
Institute of Philosophy
D-01062 Dresden
Germany

Prof. Dr. Dr. Bernhard Irrgang

TU Dresden
Institute of Philosophy
D-01062 Dresden
Germany

Webpage: http://tu-dresden.de/die_tu_dresden/fakultaeten/philosophische_fakultaet/iph/tph/irrgang/irrg
Email: bernhard.irrgang@tu-dresden.de

Prof. Dr. Kohji Ishihara

Department of History and Philosophy of Science
The University of Tokyo
3-8-1, Komaba, Meguro-ku
Tokyo 153-8902
Japan

Webpage: http://hps.c.u-tokyo.ac.jp/staff/_data/ishihara_kohji/index_en.php
Email: cishi08@mail.ecc.u-tokyo.ac.jp

Dr. Jörg Jewanski

Musikgeschichte, Musikwissenschaft
Musikhochschule Münster
Fachbereich 15 der Westfälischen Wilhelms-Universität Münster
Ludgeriplatz 1
D-48151 Münster
Germany

Webpage: http://www.uni-muenster.de/Musikhochschule/kollegium_musikgeschichte.html
Email: jewanski@gmx.de

Maika Nakao, M.A.

The University of Tokyo Center for Philosophy
3-8-1, Komaba, Meguro-ku
Tokyo 153-8902
Japan

Webpage: http://utcp.c.u-tokyo.ac.jp/members/data/nakao_maika/index_en.php
Email: goa.maika@gmail.com

Kerstin Palatini, M.A.

Anhalt University of Applied Sciences
Department of Computer Science and Languages
Postfach 1458
D-06354 Köthen (Anhalt)
Germany

Prof. Dr. Jorge Solis

Faculty of Health, Science and Technology
Institute for Engineering and Physics
Karlstad University
Universitetsgatan 2
651 88 Karlstad
Sweden

Waseda University, Research Institute of Science and Engineering
3-4-1 Ookubo, Shinjuku-ku
Tokyo 169-8555
Japan

Webpages: http://www.kau.se/en/avdelningen-for-fysik-och-elektroteknik/ee/staff/solis
http://www.takanishi.mech.waseda.ac.jp/top/solis/index.htm
Emails: jorge.solis@kau.se
solis@ieee.org

Prof. Dr. Atsuo Takanishi

Department of Modern Mechanical Engineering, Waseda University
2-2, Wakamatsu-cho, Shinjuku-ku
Tokyo 162-8480
Japan

Webpage: http://www.takanishi.mech.waseda.ac.jp
Email: contact@takanishi.mech.waseda.ac.jp

Manja Unger-Büttner, M.A.

TU Dresden
Institute of Philosophy
D-01062 Dresden
Germany

Webpage: http://www.designethik.de
Email: unger@designethik.de

Prof. Dr. Dr. Walther Ch. Zimmerli

Humboldt-Universität zu Berlin
Unter den Linden 6
D-10099 Berlin
Germany

Collegium Helveticum
Semper-Sternwarte
Schmelzbergstrasse 25
CH-8092 Zürich
Switzerland

Dresden Philosophy of Technology Studies
Dresdner Studien zur Philosophie der Technologie

Edited by/Herausgegeben von Bernhard Irrgang

Vol./Bd. 1 Bernhard Irrgang: Technologietransfer transkulturell. Komparative Hermeneutik von Technik in Europa, Indien und China. 2006.

Vol./Bd. 2 Bernhard Irrgang / Sybille Winter (Hrsg.): Modernität und kulturelle Identität. Konkretisierungen transkultureller Technikhermeneutik im südlichen Lateinamerika. 2007.

Vol./Bd. 3 Lars Leidl / David Pinzer (Hrsg.): Technikhermeneutik. Technik zwischen Verstehen und Gestalten. 2010.

Vol./Bd. 4 Arun Kumar Tripathi (ed.): Bernhard Irrgang: Critics of Technological Lifeworld. Collection of Philosophical Essays. 2011.

Vol./Bd. 5 Michael Funk / Bernhard Irrgang (eds.): Robotics in Germany and Japan. Philosophical and Technical Perspectives. 2014.

www.peterlang.com

www.ingramcontent.com/pod-product-compliance
Ingram Content Group UK Ltd.
Pitfield, Milton Keynes, MK11 3LW, UK
UKHW041438190426
11946UKWH00021B/13

9 783631 620717